Tire Forensic Investigation

Analyzing Tire Failure

Other SAE titles of interest:

Tire and Vehicle Dynamics, Second Edition
By Hans B. Pacejka
(Product Code: R-372)

**The Racing & High-Performance Tire:
Using the Tires to Tune for Grip and Balance**
By Paul Haney
(Product Code: R-351)

Car Suspension and Handling, Fourth Edition
By Donald Bastow, Geoffrey Howard, and John P. Whitehead
(Product Code: R-318)

Tires, Suspension and Handling, Second Edition
By John C. Dixon
(Product Code: R-168)

Fundamentals of Vehicle Dynamics
By Thomas D. Gillespie
(Product Code: R-114)

Road Vehicle Dynamics
By Rao Dukkipati, Jian Pang, Mohamad Qatu,
Gang Sheng, and Shuguang Zuo
(Product Code: R-366)

For more information or to order a book, contact SAE International at
400 Commonwealth Drive, Warrendale, PA 15096-0001;
phone (724) 776-4970; fax (724) 776-0790;
e-mail CustomerService@sae.org;
website http://store.sae.org.

Tire Forensic Investigation

Analyzing Tire Failure

Thomas R. Giapponi

Warrendale, Pa.

All rights reserved. No part of this publication may be reproduced, stored in a retrieval system, or transmitted, in any form or by any means, electronic, mechanical, photocopying, recording, or otherwise, without the prior written permission of SAE.

For permission and licensing requests, contact:

SAE Permissions
400 Commonwealth Drive
Warrendale, PA 15096-0001 USA
E-mail: permissions@sae.org
Tel: 724-772-4028
Fax: 724-772-4891

Library of Congress Cataloging-in-Publication Data

Giapponi, Thomas R.
  Tire forensic investigation : analyzing tire failure / Thomas R. Giapponi.
     p. cm.
  Includes bibliographical references and index.
  ISBN 978-0-7680-1955-1
  1. Automobiles--Tires--Testing. 2. Machine parts--Failures. 3. Forensic engineering. I. Title.
  TL270.G53 2008
  629.28'26--dc22   2008010119

SAE International
400 Commonwealth Drive
Warrendale, PA 15096-0001 USA
E-mail: CustomerService@sae.org
Tel: 877-606-7323 (inside USA and Canada)
        724-776-4970 (outside USA)
Fax: 724-776-1615

Copyright © 2008   SAE International

ISBN 978-0-7680-1955-1

SAE Order No. R-387

Printed in the United States of America.

*To my daughter, who inspires me,
and to my wife, who encouraged me,
not only in life, but to actually sit down and write this book.*

Acknowledgments

I would like to thank all those people within SAE International who aided me in writing this book, including all those who did so anonymously through the SAE peer review process.

In addition, I would like to give special thanks to Dr. William Van Ooij from the University of Cincinnati for his review of Section 9.1 of Chapter 9, and to Dr. James Rancourt of Polymer Solutions Inc. of Blacksburg, Virginia, for his guidance in the limited amount of chemistry that is included in the book.

Without the help of all these individuals, this book would be less than it is.

Contents

Foreword xv
Preface xix

Chapter 1 Belt Separation 1
    1.1 Belt-to-Belt Separation (or Belt Separation)—
          Discussion 1
    1.2 Belt Separation—Crack Initiation and Propagation 3
    1.3 Root Causes of and Contributors to Belt Separation 6
          1.3.1   Over-Deflection 6
          1.3.2   High Speed 7
          1.3.3   High Ambient and Pavement Temperatures 8
          1.3.4   Road Hazards (Impacts) 8
          1.3.5   Ozone Deterioration 9
          1.3.6   Physiological Damage 9
          1.3.7   Unrepaired or Improperly Repaired Punctures ... 10
          1.3.8   Improper Tire Maintenance 10
          1.3.9   Vehicle-Related Conditions 11
          1.3.10 Mounting and/or Demounting Damage 11
          1.3.11 Poor Storage of Tires 11
          1.3.12 Cuts, Snags, Gouges, Tears, and Abrasions 12
          1.3.13 Penetrations 12
          1.3.14 Manufacturing and Design Conditions 13

Chapter 2	Belt Separation Identification	15
	2.1 Tread and Belt Are Detached	15
	2.1.1 Belt Detachment Characteristics	19
	2.1.2 Belt Stock Degradation	21
	2.1.3 Road Rash	21
	2.2 Intact Top Belt and Tread	23
	2.3 Accelerated Wear—Underlying Separation	29
	2.4 Belt Separation—Additional Notes	33
Chapter 3	Other Types of Belt Separation	35
	3.1 Belt Edge Separation	35
	3.2 Incipient Belt Edge Separation	37
	3.3 Atypical Belt Separation	37
Chapter 4	Identification of Causes and Contributors to Belt Separation	39
	4.1 Punctures	39
	4.1.1 Over-Deflection	46
	4.1.2 Intra-Carcass Pressurization	47
	4.1.3 Water and Salt Corrosion	53
	4.1.4 Breakage of Belt Wires	55
	4.2 Over-Deflection	56
	4.2.1 Observation of the Compression Groove	56
	4.2.2 Wrinkling of the Innerliner	58
	4.2.3 Innerliner Color or Discoloration	58
	4.2.4 Exterior Sidewall Contact with the Road Surface	59
	4.2.5 Observation of the Tread Shoulders	60
	4.2.6 Wheel Weight Clip Mark Depth	61
	4.2.7 Shifting or Chattering of the Balance Weight Mark	63

	4.2.8	Interior (Liner Side) of the Bead Toe—Creased or Cracked	64
	4.2.9	Bead Face Abrasion or a Circumferential Line in the Bead Face	66
	4.2.10	Creasing of the Base Radii of the Tread Grooves	66
4.3	Penetrations		67
	4.3.1	Tread Attached to the Casing	67
	4.3.2	Tread and Top Belt Detached	70
4.4	Impacts		73
	4.4.1	Identification of Road Hazard Impacts	76
	4.4.2	Identification of Belt Separation Due to Impact	81
4.5	Ozone Deterioration		88
4.6	Mounting and/or Demounting Damage		93
4.7	Physiological Aging		98
	4.7.1	Durometer	98
	4.7.2	Appearance and Feel	99
	4.7.3	Spot Ozone Damage	100
	4.7.4	Belt Tearing	101
4.8	Snags, Gouges, Cuts, Tears, and Abrasions		102
4.9	Cutting and Chipping		104
4.10	Poor Tire Storage and Improper Tire Maintenance		106
	4.10.1	Poor Tire Storage	106
	4.10.2	Improper Tire Maintenance	107
4.11	High Speed and High Ambient and/or Pavement Temperatures		108
	4.11.1	High Speed	108
	4.11.2	High Ambient and/or Pavement Temperatures	109
4.12	Vehicle-Related Conditions		109

Chapter 5	Identification of Non-Belt Separations		111
	5.1	Tread Separation	111
	5.2	Bead Area Separation	112
		5.2.1 Lower Sidewall Compound Separation Off the Plies	114
		5.2.2 Separation Between the Ply Turn-Up(s) or Turn-Down(s)	114
		5.2.3 Separation of the Bead Wires or Bundle from the Surrounding Plies	115
		5.2.4 Separation Between the Steel or Fabric Chipper or Chafer	115
		5.2.5 Bead Breaks	116
	5.3	Sidewall Separation	117
		5.3.1 Separation Between the Plies	117
		5.3.2 Separation Between the Sidewall Compound Components	118
		5.3.3 Separation Between the Ply and the Sidewall Compound	118
Chapter 6	Identification of Various Tire Conditions		119
	6.1	Run-Flat Damage	119
	6.2	Chemical Damage to the Tread and Sidewall	124
	6.3	Non-Ozone-Related Cracking, Indentations, and Bulges	125
		6.3.1 Cracking	125
		6.3.2 Indentations	126
		6.3.3 Bulges	129
	6.4	Identification of Innerliner Conditions	130
		6.4.1 Appearance of a Lap Splice	130
		6.4.2 Appearance of a Butt Splice	131
		6.4.3 Liner Tags	132

	6.4.4	Liner Openings 132
	6.4.5	Ply Cord Shadowing in the Liner 134

Chapter 7   Identification and Significance of Balance Weight Marks 137

Chapter 8   Location of the Tire on a Vehicle 145
    8.1   Outboard Side Versus Inboard Side 145
    8.2   Rear Position Versus Front Position 147
        8.2.1   Rear Position 147
        8.2.2   Front Position 148
    8.3   Left Side Versus Right Side of the Vehicle 149

Chapter 9   Addressing Several Failure Theories 151
    9.1   Brassy Wire Failure 151
        9.1.1   No Bonding Between the Brass Laminate and the Belt Compound 152
        9.1.2   Partial Bonding Between the Brass Laminate and the Belt Compound 152
        9.1.3   Proper Bonding and Brassy Wire Appearance 153
    9.2   Manufacturing Imprints—"Liner Marks" 154
    9.3   Nylon Overlay 156

Chapter 10   Visual and Tactile Nondestructive Tire Investigation Techniques 159
    10.1   Basic Inspection Process 159
    10.2   Marking the Tire for Inspection 161
    10.3   Examination Process—Notes and Photographs 163
    10.4   Tactile and Visual Inspection of the Tire 165
        10.4.1   Serial Side Sidewall 165

		10.4.2	Beads 166
		10.4.3	Opposite Serial Side Sidewall 167
		10.4.4	Tread 167
		10.4.5	Belts 168
	10.5	Inspection Procedure—Demounting a Tire from the Wheel 169	
	10.6	Wheel Inspection 170	
	10.7	Matching the Wheel to the Tire 174	
	10.8	Identifying Multiple Past Tire Balances 176	
	10.9	Photography 177	

Appendix A	References 179
Appendix B	Terms 185
Appendix C	Compression Groove 197
Appendix D	Run-Flat Sequence 203
Appendix E	Shell Rating Scale for Ozone Deterioration 209

Index 211

About the Author 217

Foreword

Tire forensics is the methodical analysis of failed tires in the pursuit of and the identification of the cause(s) or root cause(s) of the disablement of a tire. A combination of science, experience, and some art goes into the research and analysis of a tire failure. By using the laws of physics, math, chemistry, and engineering, mixed with the expert's real-world tire background and experience in the design, testing, and tire development and manufacturing processes, tire forensic experts determine the most likely events that led up to and caused the tire to fail.

Unlike a forensic pathologist analyzing a body in a criminal case, a failed tire in a civil suit typically is considered evidence that cannot be dissected and destructively analyzed without agreement by all sides. The analysis by the tire expert also may not occur until years later, with a chain of custody that may or may not be tidy and with storage conditions that can be less than optimal. Given these conditions, the forensic tire expert's background, knowledge, and ability to determine pre-, during, or post-accident damage can be crucial to determining failure causation.

The same sidewall that will cut and tear during a curb scuff event can impact a road hazard to bend the wheel flange backwards and show relatively little damage on the exterior of the tire. It takes a trained eye working in a systematic fashion to find the unusual detail that leads to the root cause contributors to the failure. I say "root cause" because the type of failure should not be the end result of the investigator's work. It usually is the beginning. The goal of the tire forensic expert is to keep digging backwards, looking for root cause(s) and putting as much of the tire story together as possible.

In this book, I cover the many ways that a tire can fail and how to identify that failure. However, I will not be going into anything but minor depth in several sections on defectively manufactured tires. It is my opinion that an expert tire forensic investigator looking into manufacturing or design defects requires knowledge of not only tire failure mechanics, but also a solid grounding in several areas such as tire manufacturing, tire mechanics, testing, and tire design, as well as some familiarity with quality control parameters. The total range of parameters, how all the various pieces are supposed to fit and operate together, and what are the correct or incorrect manufacturing details in a tire will contain subtleties and nuances; thus, at times, only experience can dictate the outcome. These subtleties will make each tire its own case with particular circumstances and therefore will be covered only generally here. However, while the tire forensic expert is performing the inspection, he or she must always examine the tire with an open mind, looking for all possibilities of failure modes, including an improperly manufactured tire or a badly designed tire.

The terms "accident sequence," "pre-accident," and "post-accident" are used frequently in this book. Most tire failures do not result in vehicle damage, collisions, rollovers, and so forth. However, litigation-related tire failures tend to involve at least some vehicle damage. By using these terms, I am relating the tire conditions that are seen to the sequence of events after tire disablement, whether or not vehicle damage or an accident has occurred.

In nearly all of the text, I discuss passenger car and light truck tires with two-belt systems. There is passing mention of other belting structures, as well as medium/heavy truck tires. However, when the #1 or #2 belt is mentioned specifically, it is in the context of a two-belt tire.

In this book, when I mention root causes, it is not with the intent to gloss over the fact that, by definition, there cannot be multiple root causes. However, in tire forensics, considering damage factors to the tire during the accident sequence, sullied chains of custody, and other factors, it will not be unusual for the expert to opine that more than one reason caused the tire disablement and to determine that one and only one cause is simply not possible.

This book will not cover the following items:

- An overview of general tire basics or basic tire mechanics is not included. The list of references in Appendix A covers basic tire mechanics in depth. I assume the reader already has a basic understanding of the tire. This book provides the tire expert a compilation of the latest references on various subjects and serves as a reference itself.

- All possibilities of failure combinations or of what can be seen or photographed in the examination of a tire are not included. There can be details that can merge to bring out the correct causation, and at times, that subtle grouping and the nuances that come with it will come from the expert's background.

- The root causes of tire failure can involve chemistry and chemical forensic analysis of the tire. If this kind of analysis and information is required, then expert advice in those areas should be sought. This type of analysis is beyond the scope of this book.

- The photographs within this book are not in any way meant to cover the range of possibilities of the various appearances of tire forensic examination points. They are examples intended only to help the reader with the discussion topic at hand.

What will be covered in this book are the methodical, physical, visual, and tactile examination of failed tires, a discussion and identification of the various failure modes for passenger car and light truck-type tires, and how to determine some of the past history of a tire. Interspersed among various factors, I will share what general wisdom I have learned in 30 years in the tire industry.

My apologies to anyone who is unfamiliar with the terms used in this book because those terms tend to vary among those of us from different tire backgrounds. Appendix B provides definitions of terms as an aid to readers.

Finally, please read all the footnotes. These important footnotes add information to the text or indicate exceptions; however, they serve only as guide posts and are not meant to be all encompassing in all situations.

Although specific page numbers are listed with the cited references scattered throughout this book, these references provide a great deal of general support and knowledge to tire forensics and tire failure analysis.

Preface

Many years ago when I was a young tire engineer, an R&D bias medium/heavy truck tire was on the table for examination. A more senior engineer had already looked at the tire and proclaimed it an over-deflected tire (which it was). The lower sidewall turn-ups for approximately 30 cm (12 in.) lay open, and when I asked why the tire came apart, over-deflection was the answer. I later examined the separated area carefully with a scalpel, peeling away pieces as one would peel an onion. I had found the #1 and #2 turn-ups were folded back onto themselves. Cutting the tire circumferentially in the lower sidewall 360 degrees revealed that while the tire had small turn-up separations at various locations, it had folded back turn-ups only in the blown-out area. This discovery never left me. As an engineer, I find that the enjoyment of determining why something comes apart is the same as building something from scratch.

Photo courtesy of Microsoft® Clip Art.

I hope this book opens the possibility of discovery for you and in a real sense provides the technical background for that discovery.

CHAPTER 1

Belt Separation

1.1 Belt-to-Belt Separation (or Belt Separation)— Discussion

Modern radial tires[1-1] seldom fail [Ref. 1.1, Chapter 15, p. 615], and even fewer cause crashes. However, when tires do fail, the most typical failure mode is a belt separation. Therefore, the details of the identification and cause(s) of a belt separation are critical and are covered in this first chapter.

A belt separation occurs when one belt is no longer bonded to another belt, and that detachment was not present after curing.[1-2] A belt separation typically starts from a cohesive failure within the rubber, forming a very small "crack."[1-3] This process is called crack initiation, and it occurs between the working belts, typically beginning at the #2 belt edge.[1-4] The movement of the separation as it enlarges is called "crack propagation." "Crack growth" typically begins along the belt edge

---

1-1   The phrase "modern radial tires" has many different definitions. Although the first radial tire patent belongs to Michelin in 1946, it generally is considered that the radial tire era in the United States began in the 1970s. For the sake of clarity, when I refer to the modern radial tire, I am considering only the mid-1990s forward.

1-2   If the separation was there immediately after the tire was cured, that area contains a curing defect of some type rather than a belt separation. Note that the terms "curing" and "vulcanization" can be used interchangeably.

1-3   Reference 1.3 indicates the progression of this cracking. While this report was made for a specific tire investigation, the belt separation shown in Appendix A of this reference is typical of most belt-to-belt separations.

1-4   Most separations begin off the #2 belt edge, but it is not atypical to see a separation start off the #1 belt edge.

and then moves inward toward the centerline of the belt, ultimately forming a parabolic shape (Fig. 1.1) that is wider on the belt edge with the apex toward the center of the tire [Ref. 1.2, pp. 18–20]. A belt package (Fig. 1.2) technically is considered to have a separation when that separation is visible on holography or shearography. However, using this pure technical definition leads to some very small areas to call separations, and those separations may continue at the belt edge (see Section 3.2 of Chapter 3 regarding incipient belt edge separations) or other areas of the tire for the life of the tire, never causing a tire disablement. Note that the U.S. Department of Transportation (DOT) does not use shearography to define separation in its regulatory testing.[1-5]

Fig. 1.1 Typical parabolic (crescent-shaped) form of a belt separation.

A more typical identification of a belt separation is made when the small rubber cracks widen to join each other, and the separation becomes approximately 5 mm (0.2 in.) or more in width and length. In most cases, an examination of the tire indicates a large separation, and there is no quibbling over a few tenths of an inch.

[1-5] U.S. DOT testing in FMVSS 139 (Federal Motor Vehicle Safety Standards) does not use shearography for determination of belt separation. The test uses visual identification. "S6.2.2 Performance requirements. When the tire is tested in accordance with S6.2.1: (a) There shall be no visual evidence of tread, sidewall, ply, cord, innerliner, belt or bead separation, chunking, open splices, cracking, or broken cords."

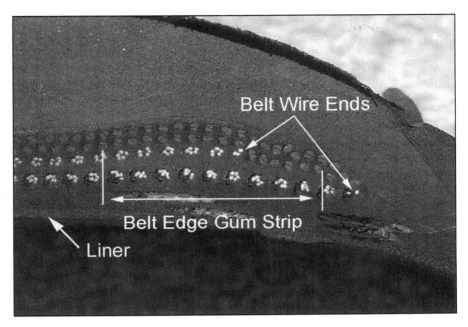

Fig. 1.2 *Tire section view of the tread shoulder.*

The cause(s) for the crack initiation and propagation can be singular, but generally there are secondary or more contributors to the failure. These are discussed later in this chapter.

1.2 Belt Separation—Crack Initiation and Propagation

A separation between the working belts begins with crack initiation typically within the belt skim stock (compound) or belt edge gum strip,[1-6] or the belt wire ends in the shoulder area of the tire. (These latter two are indicated in Fig. 1.2.) Depending on the cause(s) of the initiation of the crack, the integrity of the tire at the time of the initiation, and environmental and vehicle factors, propagation and growth can follow. The separation growth as discussed previously progresses along the belt edge (usually encompassing the belt edge gum strip) and then progresses inward toward the centerline. Crack propagation through the belt edge gum strip from the belt edge to the inward edge of the gum strip[1-7]

1-6 Not all tires are designed with a belt edge gum strip.

1-7 The inward side of the belt edge gum strip is that side closest to the centerline of the tire section.

Fig. 1.3 *Belt separation at the belt edge, as viewed from the bottom of the #2 belt in a detached tread and belt.*

(Fig. 1.3) can vary from slow to rapid, depending on the parameters causing its growth in the first place (e.g., the energy release rate). However, within the belt edge gum strip area (Fig. 1.2), the structural design of the tire and compounds are such that crack propagation and growth can be inhibited. However, once crack growth has progressed to the inward edge of the belt edge gum strip, there is a much more rapid progression of the separation [Ref. 1.4, p. 18] toward and beyond the centerline of the tire. The now loose (i.e., separated) tread and #2 belt combination is not secured to the #1 belt, and as the separation grows, the centrifugal force alone will begin to tear one belt from another.[1-8] As this parabola[1-9] appearing separation grows to or beyond the centerline, the forces on the tire may create another parabola-like separation growing inward from the opposite side. This second parabolic growth, if it occurs, will be much smaller

---

[1-8] Although centrifugal forces are always acting on the spinning tire and are trying to pull it apart, the intact tire can resist those forces. Note that excessive speed either above the speed ratings for the tire or a freely spinning tire that is not in contact with the ground can come apart, even with intact belts [Ref. 1.5, p. 8].

[1-9] This parabola also is called crescent or thumbnail shaped.

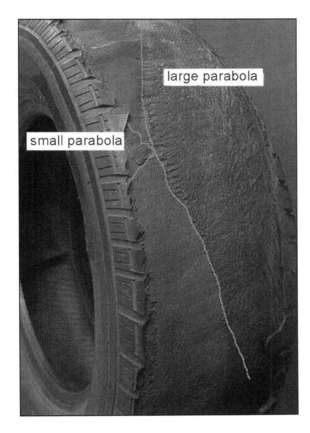

Fig. 1.4 Large and small parabolic separations joined across the casing.

than the initial separation (Fig. 1.4) but eventually will join the initial parabola, such that the #2 belt and tread now are separated 100% across the tread. From this point onward, it will be only a matter of time and speed before there is a tread and belt detachment.

Somewhere in the preceding process, for a single separation, prior to the tread and #2 belt partial or complete detachment from the #1 belt, the separation will reach a point where there will be an input into the axle on a once-per-revolution cycle. This vibration can be followed by an increased noise level coming from the tire.[1-10] The ability to hear or feel these vibrations and noises through the vehicle depends on the following:

[1-10] At 100 kph (60 mph), a once-per-revolution "thump," depending on the size of the tire, will be approximately 12 "thumps" per second (12 Hz).

- The driver or passengers being aware of a change in the operating characteristics of the vehicle.
- The driver or passengers becoming aware of a change in the sound level.
- The rapidity of the growth of the separation.
- The ability of the vehicle to mask these events.

1.3 Root Causes of and Contributors to Belt Separation

Possible root causes or contributors to belt separation and growth are as follows (in no particular order): over-deflection, high speed, high ambient and/or pavement temperatures, road hazards, ozone deterioration (also called weather checking), oxidative deterioration, unrepaired or improperly repaired punctures, improper tire maintenance, vehicle-related conditions, mounting and/or demounting damage, poor tire storage, cuts, snags, gouges, tears, abrasions, penetrations, and manufacturing and design conditions.

This section explains the definitions and scope of the preceding items. Chapter 4 discusses how to identify these causes or contributors during a tire examination and/or to provide other details during that examination.

1.3.1 Over-Deflection

Normal deflection (i.e., design deflection) in a tire is a measurement determined by the distance from the tread centerline to the center point of the wheel (or hub) of the inflated but unloaded tire, minus the distance from the same location points when the tire is loaded, matching the Tire and Rim Association (TRA)[1-11] design pressure and load. When the deflection of a tire is greater than the design deflection, the tire is said to be over-deflected.

Over-deflection is caused by under-inflation[1-12] or overload, or a combination of both. Figure 1.5 indicates "d" as the deflection between the unloaded and loaded

---

[1-11] See Ref. 1.6. In the United States, the Tire and Rim Association (TRA) standardizes loads and air pressure.

[1-12] National Highway Traffic Safety Administration (NHTSA) Tire Pressure Special Study, February 2001, Table 18, and Conclusion. http://www-nrd.nhtsa.dot.gov/pdf/nrd-01/esv/esv18/CD/Files/18ESV-000256.pdf. 27% to 32% of vehicles on the road have at least one tire that is under-inflated by at least 8 psi versus the specification shown on the vehicle placard.

Belt Separation

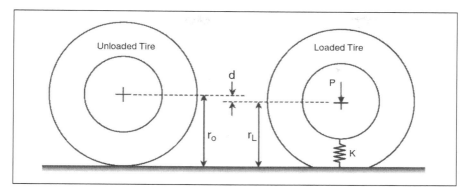

Fig. 1.5 *Schematic indicating deflection [Ref. 1.1, Chapter 15, p. 619].*

radii. Over-deflection occurs when the value of d exceeds design standards. Over-deflection causes higher strain, shear, and stress forces than normal within the tire, especially at the belt edges. Therefore, the tire works harder, which leads to higher internal temperatures in the belt edge region of the tire [Ref. 1.1, Chapter 15, pp. 618–622].

Over-deflection by itself can be the sole cause of a belt separation; however, in a tire forensic examination, this cause can be seen with other causes or contributors. Depending on the degree of over-deflection, a determination should be made regarding whether over-deflection was the root cause or a contributor of the tire failure.

1.3.2 High Speed

The ultimate speed capability of the tire basically is determined by the tire construction, the load, and the tire pressure.

Most modern radials carry a speed rating within the service description (i.e., size–load index–speed rating). Speed rating information can be found at many sites on the Internet [Chapter 4, Ref. 4.14, p. 10]. The speed rating does not mean that the tire can run for an infinite amount of time at that speed. The international regulatory and SAE tests that allow a manufacturer to put the speed code on the tire generally run from 10 to 20 minutes at that speed (ECE R30 and SAE 1561/1633). Although tire manufacturers have their own tests that typically are substantially more difficult and are run for greater periods of time than the regulatory tests, by noting the speed code, one can be sure only that the tire passed the regulatory test.

High speed by itself rarely is the primary cause of a tire failure, with the exceptions of explosive tensile-type speed failures. Due to centrifugal force,[1-13] speed can contribute to a belt separation when combined with under-inflation or overload, with heat buildup degrading the belt compound.

1.3.3 High Ambient and Pavement Temperatures

These two types of heat generation generally coincide with high amounts of sunshine and may relate directly to an increase in belt edge temperatures and increased oxidative effects on the belt compound. Excluding tire fires (mostly in medium or heavy truck [M/HT] tires), high ambient and pavement temperatures alone rarely are the sole cause of tire failures. On the other hand, high ambient temperatures usually are involved in tire failure because a tire is an engineered product that does work outdoors, is in contact with the road, and is subject to an array of environmental and damage factors that raise the internal temperatures of the tire. High ambient and pavement temperatures therefore can contribute to conditions that lead to a belt separation.

1.3.4 Road Hazards (Impacts)

Road hazards constitute all the things that lie in, on, or under (e.g., a pothole) an improved road surface or when an improved surface does not exist. Depending on the size, shape, and material of the road hazard, the vehicle speed, the tire construction, the pressure, the load, and so forth, tire component-related fractures and ruptures can be the result of impacts with road hazards. At times, an impact from a road hazard can be combined with another cause, and the impact is not the primary reason for the tire failure but the reason the initial separation propagates at a certain location. In addition, road hazards can cause belt separations, even with very little external damage noted. Cutting, chipping, and chunking are special cases of road hazard impacts and penetrations. Although these tears of the tread usually are environmental use factors (i.e., non-improved road surfaces), they are indeed small impacts and penetrations.

[1-13] Centrifugal force, $F_c = mv^2/r$, where F_c is the centrifugal force, m is the mass, v is the speed, and r is the radius. Therefore, for a given radius, the faster the speed, the larger the mass, the greater the force value.

1.3.5 Ozone Deterioration

Ozone deterioration, sometimes called weather checking,[1-14] is actually the scission of polymer chains on a molecular level. As the chains are cut by ozone and sunlight, they become visible as cracks on the exterior of the rubber. These cracks can grow with time and therefore become an environmental determinant of physical age, which may or may not be related to the chronological age. Antiozonant and anti-oxidant materials are added to almost all rubber materials in a tire to slow this type of attack on the rubber compounds.

Ozone deterioration alone rarely is determined to be a root cause of a tire failure unless a structural part of the tire is affected in some manner (e.g., the belts or the ply endings). Its main role is in determining the time-dependent events of the damaged areas, the side facing outward on the wheel, and the basic history of the tire, or as a possible contributor.

1.3.6 Physiological Damage

In this book, a discussion of tire aging will be limited to the excessive physiological damage of a tire as judged by a visual and tactile examination, rather than the chemical decomposition of the tire structure. Tire engineers and chemists have designed the tire structure to retard the effects of ozone, oxygen, and moisture attack on tire compounds. However, as time passes, ozone and oxidative changes may occur in tires due to the greater opportunity of exposure to the environment.

Today, there is no DOT test for "aging." (Such a test currently is under development.) Therefore, there is no requirement for testing tires to a specific set of aging conditions, nor is there an agreement among vehicle and tire manufacturers for a set time frame for removal or inspection of tires from a vehicle. Although several vehicle and tire manufacturers have set time limits to either remove the tires from service[1-15] or to bring them in for inspection,[1-16] the usable life of a

---

[1-14] Ozone deterioration should never be referred to as "dry rot." Technically speaking, "dry rot" is a term applied to wood and not tires; however, its connotation with tires is well known to the average person.

[1-15] For example, Continental Tire North America and Michelin Tire North America technical bulletins are available at their respective websites.

[1-16] Tire and vehicle manufacturers have varying times for inspection or removal of tires from service. (For example, as of July 2007, Ford suggests six years for removal, Toyota recommends six years for a check, and Michelin, Bridgestone/Firestone, Continental, and Cooper suggest 10 years for removal.) Some have no recommendations at this time (e.g., General Motors and Goodyear).

tire has more to do with its maintenance during that life[1-17] than the degradation of its compounds or structure.

The tire forensic expert must be knowledgeable about what constitutes the damage appearances of the exterior and interior rubber components, regardless of the chronological age of the tire. Excessive oxidative or physiological damage can be a cause or a contributor to a belt separation.

1.3.7 Unrepaired or Improperly Repaired Punctures

A puncture of a tire means an object has penetrated through the innerliner into the air chamber. The Rubber Manufacturers Association (RMA) has available online or for purchase a tire industry bulletin on the proper methods, location, area, and size of tire repairs [Ref. 1.7—wall chart]. This wall chart is the industry-accepted approach to proper repairs for light truck and passenger car tires. Improper repairs are those made to tires that do not conform to the RMA bulletin. A proper repair is a plug and a patch, and not one or the other. An unrepaired or improperly repaired puncture can lead to belt separation.

An unrepaired puncture is one in which a puncture has occurred; however, no repair has been made. This could mean the puncturing object is still caught in the tire or has been removed or ejected.

1.3.8 Improper Tire Maintenance

All major tire companies and the RMA have available proper tire maintenance bulletins that are supplied with the original equipment (OE) vehicle or generally are available to the customer at the point of sale in the replacement market. The bulletins cover the basics of proper air pressure, tire alignment, rotation, inspection, loading, and other topics. Failure to follow these procedures is considered, in the broad sense, improper tire maintenance.

Improper tire maintenance is determined by the visual aspects seen on the tire and includes many of the items described in Chapter 4 (e.g., tire cuts, tears, abrasions, over-deflection, improper storage, irregular treadwear). Identification of the various types of irregular treadwear, sometimes called malwear, that are produced by poor tire maintenance, incorrect pressure, or load versus the wear patterns, is important and is based on experience. Failure to follow proper tire maintenance

---

[1-17] Letter from D. Shea, Rubber Manufacturers Association (RMA), to J. Runge, M.D., National Highway Traffic Safety Administration (NHTSA) dated 6/10/05—NHTSA Docket 2005-21276.

procedures can be a root cause or a contributing factor for a tire failure and may lead to tire conditions that cause belt separations.

1.3.9 Vehicle-Related Conditions

Storage of the spare tire, poor vehicle alignment conditions, and poor mechanical or structural maintenance of the vehicle suspension components (including the wheels and valves) can be a root cause or a contributor leading to tire conditions that ultimately cause belt separation.

1.3.10 Mounting and/or Demounting Damage

This type of damage almost exclusively involves the tire beads.[1-18] Tires with bead tears or chunking, due to mounting or demounting of the tire, in a location from the bead heel to the bead toe can allow interior air to wick into the ply cords. The pressurized internal air (20% oxygen plus moisture, depending on the air source) now is flowing directly into the casing ply cords, having bypassed the innerliner, and is adjacent to the #1 belt. The oxygen with pressure will begin to attack the belt rubber compound from the inside out much more rapidly than if the breach of the innerliner had not occurred. This process is called intra-carcass pressurization (ICP)[1-19] and leads directly to accelerated degradation of the belt compound.

1.3.11 Poor Storage of Tires

Poor storage of tires can include the following:

- Storage of the tires in direct sunlight
- Storage of the tires in high ambient heat
- Location of the tires near electrical motors or welders

[1-18] Lower sidewall damage sometimes can occur from the equipment that unseats the beads.

[1-19] Air is always passing through the casing from the interior to the outside of the tire through the microscopic spaces in the rubber. That is why tires may lose approximately 1 psi per month from this normal process. Intra-carcass pressurization (ICP), as discussed in this book, is the accelerated form of this normal air passage.

- The tires crushed radially or circumferentially
- Location of the tires in petrochemical ingredients
- Long-term storage of the tires in a spare-tire rack
- Storage of the tires in water

Tires should be stored in dark, dry, and cool conditions, away from ozone-producing equipment.[1-20] They should be stored no more than four high in a stovepipe-type fashion (horizontally) or individually straight up and down (vertically) in tire racks.[1-21] Tires stored in poor conditions eventually will show signs of physical deterioration, including but not limited to possible rubber creasing, ozone deterioration, and increased durometer measurements. Poor storage can lead to belt separations, as well as tread and belt detachment.

1.3.12 Cuts, Snags, Gouges, Tears, and Abrasions

A cut is made with an object that slices the rubber of a tire. On the other hand, a tear occurs when the rubber is pulled apart (e.g., a tensile tear) or crushed (e.g., a compression tear). A snag occurs from a dull object that "grabs" the rubber of a tire, ripping it apart but generally leaving one end attached. One step more than a snag is a gouge, where the rubber has been removed from the tire. These conditions—depending on the depth, length, and location—can become sites for belt separation, especially when the depth brings them (e.g., a cut, snag, or gouge) directly to or near the reinforcement components of a tire. However, depending on the type of cuts, snags, and so forth, most of these items will be used in the identification of the history of the tire pre-, during, or post-accident. (See Sections 4.4.1 and 4.4.2 in Chapter 4 for a discussion of damage trails.)

1.3.13 Penetrations

Penetrations occur when objects have entered the tire but have not reached the air chamber. Therefore, by definition, penetrations have not breached the liner.

[1-20] For example, Yokohama Technical Service Bulletin 11/21/02, Tire Storage Recommendations, http://www.yokohamatire.com/pdf/tsb-112102.pdf.

[1-21] Tires eventually will take a "set" if crushed. Tires can be shipped or stored for short periods of time in non-optimum conditions but should always be stored as soon as possible in proper conditions for use.

They are not punctures but rather can cause belt separations by cutting, tearing, or drilling into and sometimes through the belt package, cutting belt cords or ply cords, and being sites for accumulation of grit and water, salt, and ozone affecting the steel belts and therefore causing crack initiation. In addition, penetrations and cuts, snags, gouges, tears, and cutting or chipping describe something of the usage environment for the tire. Penetrations alone can degrade the belt package, leading to belt separations; therefore, penetrations can be a root cause of or a contributor to belt separations.

1.3.14 Manufacturing and Design Conditions

A determination of the potential causes for belt separation arising from these conditions requires expert knowledge of how they can happen in a factory or a design group, what the field or testing results would be from those conditions, and what constitutes a proper tire design. Therefore, it is critical that the expert's career include experience in a factory and/or a design, testing, or quality group at some point.

Mal-manufacturing or mal-design determination should be made on a case-by-case basis. A manufacturing or design defect rarely is so gross that it is completely obvious to all experts. Generally, manufacturing or design defects by themselves are rare due to the enormous quantity of testing and repeat testing of tires and their components.

In addition to the finding of a defect, that defect must be linked to the causation for the separation. An out-of-tolerance condition in a tire, if not connected to the failure, is not a defect that can be used to show causation for that defect.

"Defects" also can mean different things to different people involved in the tire industry or persons involved in the litigation [Ref. 1.1, pp. 637–638]. For instance, to a tire salesperson, a defect can mean that the tire label has been placed crooked on the tread. To a tire quality person, any "off color" of a whitewall can be a defect. Therefore, clarifying the type of defect,[1-22] indicating its causation and how that defect can occur, is part of the tire forensic job in litigation.

To review all the various nuances and technologies that might or might not lead to the conclusion of mal-manufacturing or mal-design is beyond the scope of this book. Note that a tire is not defectively designed because it does not incorporate the latest technologies.

[1-22] There are two basic classes of defects: those that affect the structural integrity of the tire, and those that do not. These two examples do not affect tire integrity.

CHAPTER 2

Belt Separation Identification

2.1 Tread and Belt Are Detached

As previously discussed, the shape of the belt separation that causes a detachment of the tread and belt often is parabolic in nature. The identification of the parabola and the extent of the separation are made from a visual check. Belt separations can have any or all of the following visual aspects in the belt compound:

- "Short" belt stock tears (Figs. 2.1 through 2.3)
- Polishing of the belt stock (Figs. 2.1 and 2.3)
- Reversion of the belt stock
- Bluing of the belt stock[2-1] (Figs. 2.4 and 2.5)
- Darker colors to the belt stock than the surrounding compound

The degradation of the belt compound properties within the separated area can be detected visually as a change in general appearance versus other areas of the tread belt detachment.

[2-1] Bluing of the rubber is a sign of heat buildup. This characteristic in the rubber can occur at a belt separation or away from the separation point and may be present in compounds other than the belt. This feature gets its name from the color of severely heat-affected rubber; however, it can be various shades of blue, purple, or dark green and can appear externally (Fig. 2.3) or internally in nearly all rubber compounds.

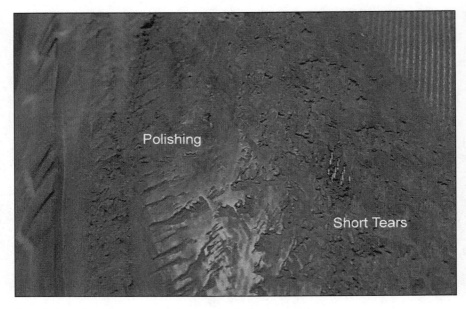

Fig. 2.1 Belt separation, indicating polishing and short tears.

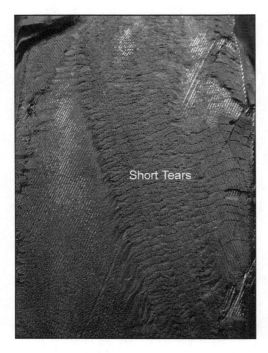

Fig. 2.2 Belt separation, indicated by short tears.

Belt Separation Identification

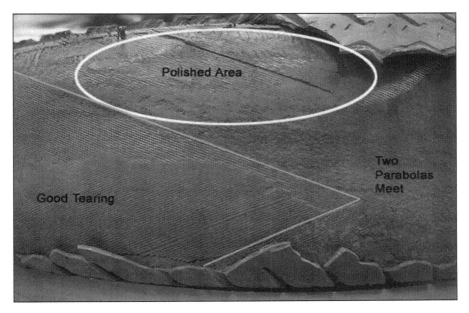

Fig. 2.3 Parabolic belt separation, identifying polishing versus good tearing.

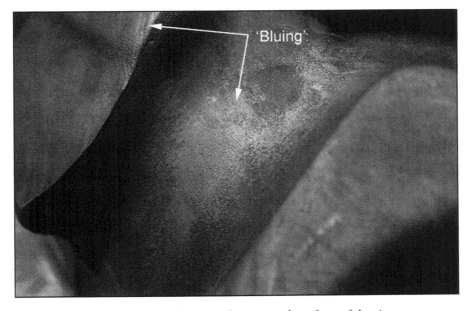

Fig. 2.4 Bluing between the lugs on the external surface of the tire.

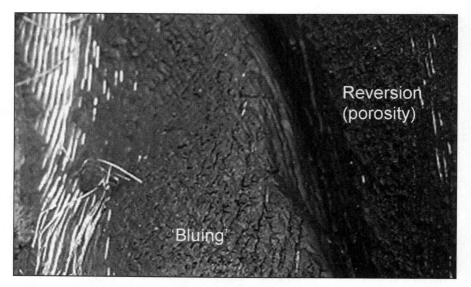

Fig. 2.5 Belt separation with bluing of the belt stock.

A belt separation may show some or all of the following characteristics:

- Texture (Figs. 2.1 and 2.2), that is, the macro- and microstructure is rough
- Color bluing (Figs. 2.4 and 2.5)
- Luminosity (reflections off a light source, Fig. 2.1)
- Smoothly worn areas (polishing, Figs. 2.1 and 2.3)
- Short tears in the rubber versus long tears visible in non-degraded belt rubber (Fig. 2.3)
- A brittle appearance or feel versus the surrounding non-belt separated areas

The analysis of the fractography of the belt separation (that being the start/stop flaps, the stop/start marks, the yaw marks, and so forth) should be a part of the repertoire of an expert's background. Both of these topics are covered by Daws in two articles [Ref. 1.2 in Chapter 1 and Ref. 2.1].

When analyzing the belt surface of either the casing or the detached belt, keep in mind the time-dependent events. For instance, rounded edges of belt fatigue cracks and polishing take more time to develop, whereas sharp-edged belt cracks can indicate recent tearing.

2.1.1 Belt Detachment Characteristics[2-2]

There are several characteristics of belt detachment that are quite normal but also are useful in the forensic analysis of a belt separation.

- **Thick/thin belt rubber**—"Thick/thin" relates to the appearance as seen on the top side of the #1 belt that is attached to the casing. However, the mirror image should appear on the bottom side of the #2 belt. The thick/thin appearance presents itself circumferentially as a visible difference in the belt rubber thickness on one half of the belt surface versus the other half. The split from thick rubber to thin rubber occurs at approximately the centerline of the belt and traverses circumferentially around the tire (Fig. 2.6). This thick/thin phenomenon has to do with the complex tearing of the tread radially across a convex radius (the tread radius) and the circumferential tearing on the radius

Fig. 2.6 Thick/thin rubber.

2-2 The general characteristics of the #2 belt detachment and the resulting marks on the #1 belt of the separation can be found in some detail in Ref. 1.2 in Chapter 1 and Ref. 2.1.

of the "round" tire. On which side of the belt the thick or thin side is located and the leading and trailing edges of the separation all can help in determining the belt separation beginnings and endings and potentially the direction of rotation of the tire. The thick/thin characteristic is viewed easily and is a condition seen commonly in tires with full or partial tread and belt detachments. It occurs with the peeling of the #2 belt off the #1 belt in field or laboratory detachments and therefore does not occur in failure types where the belts do not detach from each other. The thick-gauge rubber on the #1 belt is most of the belt rubber from both belts, whereas the thin-gauge side is the result of the belt rubber from the #1 belt staying with the detached #2 belt. Photos of this thick/thin phenomenon can be found in Ref. 2.1.

- **Transition zone**—A transition zone between the thick and thin sides exists around the circumferential centerline of the tire. The width of this transition zone varies, and it is found in all tread/belt detachments with thick and thin zones. At the transition zone, as the belt tears from one side of the thick/thin interface to the other side, it is not uncommon for the tearing to go from multi-planar to within the plane of the belt-to-belt interfaces. Due to this tearing across a plane, it also is not unusual to see smoothness to the rubber or manufacturing imprints. Because this transition zone is at the centerline, where the belt strains prior to the belt tread detachment are minimal, this area is rarely a cause of a belt separation.

- **Edge cracking**—Edge cracks are caused by crack propagation along the belt edge, within the belt edge gum strip and the belt compound. Due to the extra gauge within the belt edge gum strip, these tears, if not polished or worn down, will be quite visible. (See footnote 1-6 in Chapter 1.)

- **Stop/start marks ("beach marks")**—These marks are from the stop/start tearing and detachment of the tread and the #2 belt off the #1 belt, and they mark the area where the tear stopped and then started again. Some forensic authors retain the opinion that each stop/start mark is one revolution of the tire. Daws [see Ref. 1.2 in Chapter 1] presents data in agreement with this opinion. However, with the complexity of the tread belt tearing, I respectfully believe that each stop/start tear may not represent one revolution of the tire. Because tread and belt detachments can involve impact with the frame of the vehicle pre-accident or during the accident sequence, this effect can lead to additional stop/start tears within a revolution. Barring counting stop/start marks to determine the length of time for the tread separation, stop/start marks are quite common and, if required for the inspection, can be analyzed as part of the tearing analysis of the belt separation. Stop/start marks generally are considered good adhesion marks in a tread and detachment; however, a lack of them does not necessarily mean poor adhesion exists. Because of the nature of the complex tearing process, the length of the tread strip being torn, and impact or non-impact with the vehicle frame at times, a tread can peel a considerable distance and not have a stop/start mark.

2.1.2 Belt Stock Degradation

Many terms are used when observing a belt separation, such as those already described in this chapter. For simplicity, those terms can be reduced to "belt stock degradation" (BSD). During inspection of the tire, BSD can be used to cover all the more specific and technical terms, such as polishing, bluing, short tears, rough texture, and reversion by basically meaning the original belt compound properties appear to be degraded. Although this term generally can replace all the other terms, it is not meant to replace them technically because those other terms are technically more specific.

Of all the terms listed, reversion is the most technical and has a very specific definition to chemists and tire engineers. To chemists and tire engineers, reversion means that the rubber properties have degraded, as indicated by the negative slope on a rheometer curve. To forensic engineers, we are looking for gumminess, glassiness (i.e., embrittlement), or a porous condition [Ref. 1.1 in Chapter 1, pp. 622–623] (not to be confused with undercure) in the rubber. In a belt separation, reversion may be found in high heat and/or in highly polished areas.

For most tire inspections, this kind of highly technical identification may not be required, and using the acronym "BSD" will suffice.

2.1.3 Road Rash

An inspection of a tire casing with the #2 belt and tread detached typically yields a "clean look" at the #1 belt surface. The exceptions to this clean look are when severe accident damage is present or when the tire has not lost air and was run on an improved road surface for even a short period of time. Off-road travel for short distances after the tread and belt have detached leaves a different kind of abrasion that is rougher and with small obstacle imprints into the casing.

With road abrasion, much of the #1 belt rubber or portions of the belt wire itself can be worn away (Fig. 2.7), including at times much of the visual parabola and rubber stock (Fig. 2.8). In nontechnical terms, this abrasion of the rubber and/or the #1 belt is termed "road rash."

Fig. 2.7 Road-abraded belt wire, shown here at 15x magnification.

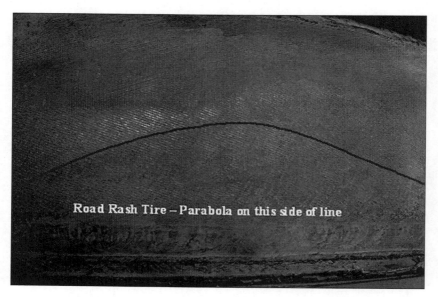

Fig. 2.8 Road-abraded tire and identification of the belt separation.

In a tire with moderate to severe road rash, much of the rubber is worn away. Therefore, a very close inspection of the #1 belt surface is needed to piece together the size and shape of the parabolic separation. However, the detached tread #2 belt bottom surface will be the mirror image of the road-abraded #1 belt top surface. This detached tread and belt will be damaged substantially less if the detachment occurred pre-accident and therefore is somewhat "clean" and valuable for forensic analysis. For this reason, saving the detached tread and belt for later analysis is important to the forensic analysis.

Improved surface road rash will appear as "tight" striations and abrasions across the #1 belt surface. At times, these will be circumferential, in the straightforward running direction, or on an angle. Angular road rash, especially in the steel belt wires, indicates that the tire made maneuver(s) after it lost the tread and belt. Simply the fact that road rash appears can be important in determining when the tread and belt detached if the vehicle left the road during the accident sequence.

2.2 Intact Top Belt and Tread

When the top belt is separated from the bottom belt but is not detached from the casing, belt separations and belt edge separations are still visible at times from the exterior and interior.

The main exterior visible characteristics of tires with a belt separation are as follows:

- **Accelerated treadwear (also called "localized treadwear") in a localized area**—[Refs. 2.2 through 2.4, and Ref. 4.1, p. 401] It can be nearly any shape but is mostly oval (Fig. 2.9).

- **Accelerated treadwear in a continuous circumferential line in either shoulder of the tire** (Fig. 2.10).[2-3]

- **Tread bulges or knots that occur in the upper sidewall of the tire adjacent to the tread**—(Fig. 2.11) If the bulges are from a separation, they will have a soft or spongy feel to them.

- **Tread cracking in the grooves or slots between the lugs or ribs of the tire that is not related to weather checking**—(Fig. 2.12) Note that tread cracking from a separation will have an underlying softness from the separation and can be felt using a tread block spreader that pushes the blocks apart so that the blocks do not feel tight to the casing.

2-3 Looking at a tire circumferentially, this would appear as a swale in the shoulder area.

Fig. 2.9 Accelerated treadwear in the shoulder from an underlying separation.

Fig. 2.10 Underlying belt separation viewed as a swale in the left shoulder.

Belt Separation Identification

Fig. 2.11 Bulge in the upper sidewall.

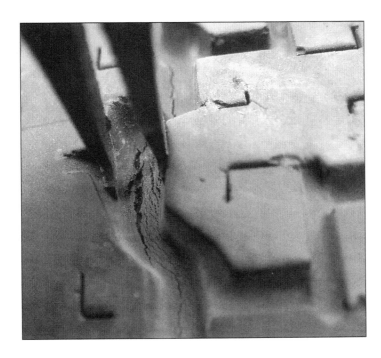

Fig. 2.12 Cracking between the lugs from underlying separation.

- **An odd appearance or shape of the tread shoulder lugs or ribs or tread surface**—For example, this might appear as a sudden change in tread radius in the radial or circumferential direction.

The main interior (liner-side) visible characteristics of tires with a belt separation are as follows:

- **Bubble(s) appearing in areas in the tread shoulder**—(Fig. 2.13) These bubbles will be softer or spongy to the feel versus non-bubbled areas in the shoulders. If there are bubbles, they will appear more or less in a circumferential line.

Fig. 2.13 Bubble in the liner shoulder.

- **Belt wire filaments visible at times and poking through the innerliner** (Fig. 2.14).

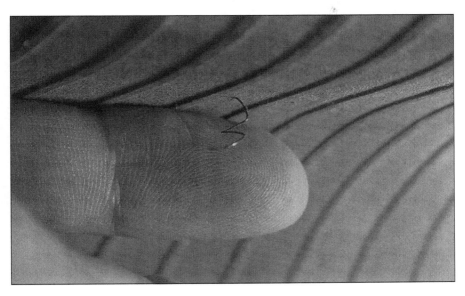

Fig. 2.14 Belt wire filament through the innerliner.

The filaments will be quite small; therefore, they are easy to miss visually. A good tactile examination of the liner surface will find them. A bare hand should be used for any interior tactile examination, but care should be exercised because these filaments can penetrate the skin!

The belt wire filaments seen through the liner are from the belt wire cables. If the belt wire filaments are broken from flex (bending), one or both ends of the filament will show a flex break. The flex breaks occur after the separation forms, and the now broken wires that also are loose from the rubber matrix can begin to migrate. The broken filaments work their way through the liner of the tire, becoming visible in the interior and at times through the tread to the surface (Fig. 2.15).

Magnification of 15x will sufficiently enlarge the ending of the wire for its proper flex break identification. Tensile breaks (Fig. 2.16) are distinctive in appearance, as are flex breaks (Fig. 2.17).

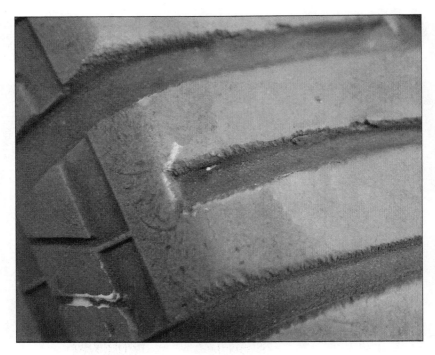

Fig. 2.15 Belt wire filament through the tread.

Fig. 2.16 Scanning electron microscope (SEM) photograph of wire breaks in tensile. (Courtesy of Polymer Solutions Inc., Blacksburg, VA.)

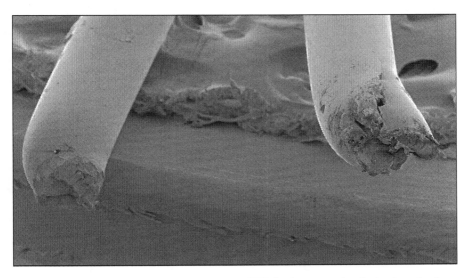

Fig. 2.17 Scanning electron microscope (SEM) photograph of wire breaks from flex. (Courtesy of Polymer Solutions Inc., Blacksburg, VA.)

Investigation should be made of non-belt separation causes of belt filament breakage, such as cut belt wires (Fig. 2.18) from penetrations, cuts, and so forth, and tensile breaks (i.e., from road hazards) and other causes of flex or abrasion breaks, that is, from vehicle maneuvers or braking after tread belt detachment.

With the tread and belt mostly intact and where there is not a clear view of the belt separation, an estimation of the length or width of the belt separation can be made by using accelerated wear areas, cracking between the lugs, looseness (softness) of the tread lugs, liner bubbles, and belt wire filaments through the liner. The full extent of the separation will be somewhat larger than the visible exterior or interior points of separation. For smaller separations that do not show externally or internally, shearography can be helpful to determine the degree of belt separation in a mostly intact tire. Expertise in reading the shearography results may be required because there typically is no baseline to use in judging the results.

2.3 Accelerated Wear—Underlying Separation

Accelerated wear is a localized worn area within the tread that is more worn than adjacent surrounding areas and that has a treadwear measurement pattern

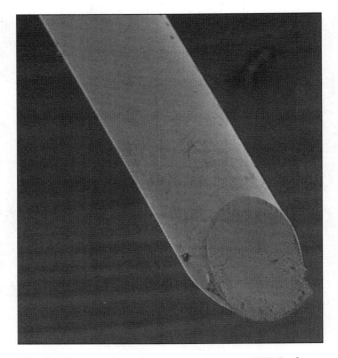

Fig. 2.18 Scanning electron microscope (SEM) photograph of cut wire. (Courtesy of Polymer Solutions Inc., Blacksburg, VA.)

of decreasing wear rate (i.e., faster wear), followed by increasing wear rate (i.e., slower wear) as one moves into, through, and out of the accelerated wear area.

Accelerated wear typically is caused when the underlying belts become loose from the structure, as in a belt separation or the tread detaching from the top belt. When the area with the loose belt structure reaches the road surface, there will be abnormal squirming of the tread, causing a higher abrasion in this area versus the areas with a tight belt structure. This is due to two phenomena: (1) when not against the road surface, the rotating tire will pull this area slightly apart from the underlying belt due to centrifugal force, and (2) the loose structure will squirm more than the surrounding tight structure. If air has become entrained in the separation area and the tire now has a bulge visible in the tread, the wear rate will be significantly faster (4 to 13 times faster) than if the separated area had no air [Ref. 2.3, p. 16]. While the outlying edges of the accelerated wear

Belt Separation Identification   31

Fig. 2.19 Circumferential accelerated wear area in the intermediate area of the tread.

area can be nearly any shape, the area most often is circular or oval in nature (Fig. 2-19).[2-4]

Accelerated wear should not be confused with irregular wear, which generally is much more widespread in the tread pattern and is not localized in nature. Furthermore, accelerated wear will have a gradual approach and exit to the adjacent areas of "normal"[2-5] treadwear, whereas irregular wear generally will not be gradual. A gradual approach is best described as the fastest wear rate that will be the center or near center of the underlying loose belt structure. From that point, as you approach any tight belt structure, the wear rate decreases (more kilometers [miles] per millimeter per 32nd of an inch). This gives a boundary that is somewhat gradual toward accelerated wear, rather than an abrupt ending point.

2-4   An exception to this shape rule occurs when the accelerated wear is located only in the shoulder and can be an inch or two wide by 360 degrees around the circumference (see Fig. 2.10). Figure 2.19 indicates the start of the swale.

2-5   Normal treadwear is the tire miles per number of millimeters per 32nds of an inch in the part of the tire that does not have a loose belt structure or non-bonded area.

When the accelerated wear area is worn such that the siping in that area starts to disappear,[2-6] that area becomes more visible and will continue to become even more clearly visible with time, until either a tread belt detachment occurs or the area finally is observed and the tire is removed from the vehicle.

Due to the various degrees of accelerated wear, the possible locations (i.e., outboard or inboard side of a vehicle), and the type of vehicle fender well, localized accelerated wear is at times either easy or difficult to spot on the vehicle.

The amount of accelerated wear present will depend on the following:

- **The size of the underlying separation**—The separation needs to reach a size such that the tread squirm in that area is sufficiently different than the surrounding area, revealing itself in the worn appearance of the tread.

- **The rapidity of the growth of the separation**—An accelerated wear area requires a slower rate of growth of the separation, allowing time for the tread-wear differences to become visible. Rapid growth separations generally do not have the necessary time for localized accelerated wear development.

- **The usage of the tire**—That is, the load, speed, or pressure.

The amount of difficulty in identifying the accelerated wear will depend on the following:

- The amount and depth of the siping

- The complexity of the tread pattern

- The remaining tread in the tire

- The experience of the examiner in determining an early accelerated wear from a malwear condition

- The size of the underlying separation

- The initial subtleness of the wear

[2-6] The "disappearance" can be a complete elimination of the sipe, or the sipe will change in shape from those same sipes in another area of the tire (see Fig. 2.20). This is the visual clue that a depth change has occurred.

Belt Separation Identification

In passenger car tires, localized accelerated wear typically is caused by an underlying separation.[2-7] By careful examination of any changes in the siping circumferentially (Fig. 2.20) or radially around the tire, accelerated treadwear that appears in a laboratory environment can be identified, even if it is subtle. In Fig. 2.20, the accelerated wear is viewed easily. However, note the siping variability, which will be the earliest indication of accelerated wear.

Fig. 2.20 Accelerated wear in a tread shoulder, with siping variability noted.

2.4 Belt Separation—Additional Notes

It is difficult to impossible to determine when a belt separation (i.e., crack initiation and some propagation) actually began because separations within the belt edge grow at varying rates, depending on the varying past and present conditions that exist in the tire (e.g., tire integrity, load, pressure). Furthermore, the crack

---

2-7   There are times, specifically in some light truck and all heavy truck tires with very stiff tread and belt packages, when accelerated wear can be caused by overlapping components. Also, non-belt-separated but non-bonded areas in the tread or belt structure may cause accelerated wear.

propagation within the belt edge gum strip area will have a much slower rate of growth than the rate when the crack penetrates beyond that area toward the centerline.

Additional clues to the rapidity of the separation lie with the amount and type of localized accelerated treadwear in this area and the degree and type of belt stock degradation seen in the underlying belt (i.e., polishing) [Ref. 1.2, p. 20].

All damage to a tire is cumulative. Tires do not heal themselves; therefore, the damage that finally resulted in the belt separation may have occurred at some time in the more distant past.

Tires that produce localized accelerated wear from underlying belt separations generally will be longer-term separations versus those without accelerated wear, which occurred fast enough that the accelerated wear did not have time to progress.

Always be aware of multiple causes and additive effects that cause belt separation. Although someone may want to know only the exact one cause of a belt separation, in the real world, the additive (i.e., contributing) damage can bring about a tire disablement. Depending on the condition of the tire at the time of the examination, the determination of why the tire ultimately failed, and assuming that contributing causes do exist, those contributing causes should be identified as well.

CHAPTER 3

Other Types of Belt Separation

3.1 Belt Edge Separation

A belt edge gum strip may be used to increase the belt compound gauge, aiding in the modulus step-down from the steel belt edges. It is one of the design fatigue growth inhibitors at the belt edge.[3-1] Because this design feature is used in most tires, a separation that begins at the edge of the belt wires eventually can involve the gum strip. Belt edge separations (BES), as opposed to the larger parabolic belt separations (Chapter 2) that cause tread and belt detachments, are circumferential, mostly linear, and located at the belt edges (Fig. 3.1). The width usually is less than 2.5 cm (1 in.), [3-2] but they can move toward the center in small parabolas around the circumference (see Fig. 1.2 in Chapter 1).

The BES can be located on one or both shoulders and can vary in its width and at times disappear entirely as the tire is rotated 360 degrees during the inspection.

Belt edge separation is identified in much the same way as the larger parabolic belt separation. However, the belt edge tearing through a non-separated belt

---

[3-1] A belt edge gum strip is a common method of accommodating the step-off of a module, but it is only one method. Some tire companies do not use a belt edge gum strip in their designs and mitigate the modulus step-off in a different manner (e.g., increased belt compound gauge).

[3-2] If the separation continues to be linear in nature and not parabolic but the width is approaching one-quarter of the width radially across the casing, then a determination must be made that it is a belt separation and not a belt edge separation. There is no bright line standard for this determination.

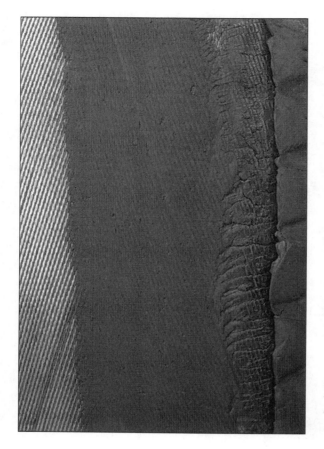

Fig. 3.1 A belt edge separation, viewed from the top of the #1 belt.

edge gum strip (Fig. 3.2) must be differentiated from that which occurs from a separated area.

Fig. 3.2 A torn belt edge gum strip—no belt edge separation.

3.2 Incipient Belt Edge Separation

Over time, steel-belted radial tires used in the proper manner[3-3] may develop very small belt edge separations at the belt wire cable ends (and at times joining from cable end to cable end). This is referred to as an incipient separation (Fig. 3.3). This incipient separation may be continuous around the circumference of the belt edge, or it may be sporadic in nature along the circumference.

Fig. 3.3 An incipient belt edge separation.

When a tire is used in the proper manner, the proper structural design and chemical composition at the belt edge are more than sufficient for the tire to run out its treadwear life, with the incipient separation remaining as such and without the belt separation growth seen in a failed tire.

3.3 Atypical Belt Separation

A typical belt separation is one that occurs between the working belts in a tire and involves the belt edge(s). Because separation growth can take several forms,

---

3-3 "Proper manner" here means within the design parameters of the tire, as dictated by the vehicle and/or tire manufacturer.

including the final tread and belt detachment from the casing, experience is the best judgment that a typical separation did not grow into an atypical type.

If you find atypical separations, a closer look at the tire structure, contamination, and/or misuse of the tire is warranted, especially if a separation in the belt package away from the belt edge is found. A separation by itself in a belt but in a non-belt edge area would be rare because the stress or strain away from the belt edge, while not zero, is substantially less than that at the edge. When it is time to put all the facts together, the type of separation can lead you down different paths.

Some examples of atypical belt separations in two-belt tires are as follows:

- The separation will start at the #2 belt edge and go over the top of the #2 belt (between the belt and the tread and without a separation between the belts).

- The separation will start at the #1 belt edge and progress under the #1 belt (between the ply and the #1 belt and without a separation between the belts).

- Belt separations are completely away from the belt edges.

- Interfacial belt separations are those separations that can occur along the plane of the individual belt compound without any sign of adhesion.[3-4] Although the naked eye usually is enough to determine interfacial tearing versus planar or multi-planar tearing, magnification may be required to determine if the belt separation indeed is exactly along a plane or within the rubber itself. Small tears within the rubber, which cannot be seen by the naked eye, can become visible with 7x or 15x magnification. Even extremely small tearing means an interfacial separation, as identified here, did not exist. Although a tire is a laminate product and is not homogenous [Ref. 3.1, p. 4, and Chapter 9, Ref. 9.5, p. 2], and laminate tearing can and does produce tearing across a single plane, true interfacial separations tend to be smooth in nature. Because tires can and do tear apart adhesively and cohesively [Chapter 9, Ref. 9.5, p. 3], experience and/or non-destructive testing and/or magnification may be needed at this point to identify the separation definitively as interfacial.

[3-4] When belts become detached, the tearing from the inside edge of one belt edge gum strip to the inside edge of the other side can tear in planar and multi-planar manners. The belt skim coat compound between the belt cables is a narrow gauge; therefore, physically there is not much room for deep tear patterns to occur.

CHAPTER 4

Identification of Causes and Contributors to Belt Separation

4.1 Punctures

In a pure sense, identification of a puncture is fairly straightforward because the definition is clear. (See Appendix B for the definition of a puncture.)

However, finding a puncture in a tire that cannot be mounted and/or aired up and put into a dip tank is difficult. Both the interior and exterior of the tire must be searched systematically [Ref. 4.1, pp. 392–393]. It is simply the nature of tire forensics that by the time a tire arrives into litigation, a puncture, if one exists (Fig. 4.1), usually is not the unrepaired nail puncture with the nail still in the tire (Fig. 4.2) but rather is the more difficult type of puncture to locate.

Although a tire can have many penetrations into the tread and belts, any of which can be a puncture, the liner generally has been protected from the environment, and its inspection is important to finding a puncture.[4-1] Probing a potential puncture of a tire in litigation can be problematic and, at times, not allowed. When probing is allowed, the probing must be done gently but firmly with a tool that has a

---

4-1   Inspection of the tire while it is on the wheel negates a complete inspection, which would include the liner side of the tire. Complete and thorough inspections are done with the wheel and tire completely demounted. (See Fig. 4.3 later in this chapter, indicating liner damage from a nail puncture.)

Fig. 4.1 A puncture with broken belt cables.

Fig. 4.2 An unrepaired nail puncture in the tread surface.

Identification of Causes and Contributors

dull front edge but is small enough to slide[4-2] through the width of the puncture. When you have entered the puncture or penetration with a tool and have reached the belts, the best way to proceed is to shift the tool slightly sidewise to the side toward the belts. Then while pressing into the belts and not the "hole," look at this location from the liner side of the tire. You should be able to see any slight protrusion in the liner, identifying the area you are probing, and you should be able to obtain a good visual on whether the liner is punctured.

For a puncture to be responsible for a belt separation, the puncturing object must have remained with the tire for some time (i.e., cycles), and the initial damage caused by the puncture is such that the tire does not run-flat or experience a sudden air loss.

Most puncturing objects tend to remain with the tire because rubber usually holds onto objects, even with centrifugal force trying to eject those objects. If the puncturing object is ejected soon after the initial puncture, a run-flat from this puncture could ensue rather than a belt separation.

The lower the tread non-skid (depth), the more likely the tires are to be punctured. For example, nails that are 8 mm (10/32 in.) in length can puncture a tire with 1.6 mm (2/32 in.) of remaining tread, whereas a remaining tread of 8 mm (10/32 in.) will not be punctured. Some penetrating objects can continue to drill down into the tire as the tire wears, eventually damaging the belts or puncturing the tire.

It is difficult to puncture a tire with an object when the tire has no internal pressure. However, inspection of the puncturing object (assuming it is still with the tire) can reveal whether or not the puncture occurred prior to the tire disablement or during the accident sequence. Examination of the object on the tread surface should be done in detail and with slight magnification. The ozone cracking around the puncturing object, the amount of remaining "head" of the nail or screw and so forth, its depth above or below the tread surface, and the striations through the metal object typically can make the determination that the puncture occurred prior to the accident.

When investigating the puncture on the liner side, look at the puncture hole and the amount of rubber movement around the hole. If the puncturing object was embedded in the tire for some time prior to its failure (Fig. 4.3) and had some

---

4-2   Probes normally do not "slide" along a rubber interface. Do not use any substance that would make this easier (e.g., soap). Pushing the probe into the hole may require some working back and forth with the tool. Be gentle because this is evidence being probed, and further damage to the tire should be avoided.

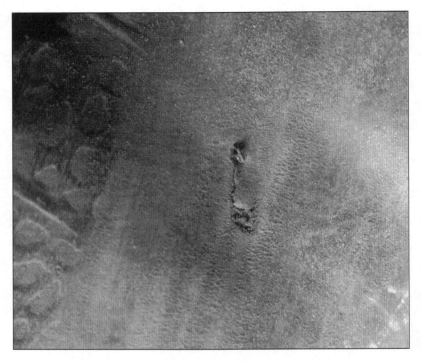

Fig. 4.3 A nail hole, as viewed from the liner side, with interior damage.

length into the interior of the tire, there is a good possibility that the hole will be wedge-shaped (i.e., larger at the opening at the liner surface and angled downward and inward toward the belts) due to the movement of the object while it was stuck into the tread (Fig. 4.4). However, if the puncturing object entered the air chamber for only a short distance, there might not be a wedge shape in the liner, even though the object may have been in the tire for some time (i.e., cycles).

A breach of the liner from an object that is held in the tire allows a continuing small amount of air out of the tire and into the casing, and allows water and possibly other external items (e.g., dirt, mud, or salt) into the "wound." In addition, depending on the size and type of the puncturing object, there could be damage to the belt system.

The following items can begin crack initiation, propagation, and growth as a result of a puncture:

Identification of Causes and Contributors 43

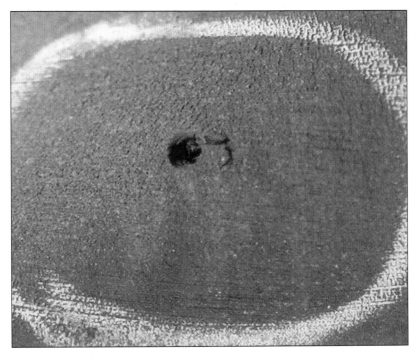

Fig. 4.4 A wedge-shaped puncture, as viewed from the liner side.

- **Air leakage from the tire**—This air leakage causes over-deflection, thereby increasing the strain, stress, and temperature in the belt/shoulder region of the tire. All of these degrade the belt compound properties, leading to crack initiation and propagation.

- **Air leakage into the casing**—This air leakage begins the process of intra-carcass pressurization (ICP) [Ref. 4.1, p. 393], which is an accelerated oxidation process within the belt structure. Intra-carcass pressurization degrades the belt compound over time.

- **Rust formation**—The entry of water and grime into the puncture can cause formation of rust at the site. Punctures by organic matter such as wood (exterior view of Fig. 4.5, and interior view of Fig. 4.6) will draw substantially more water into the puncture wound than non-organic matter (e.g., nails).

Fig. 4.5 A puncture site with rust and broken wires.

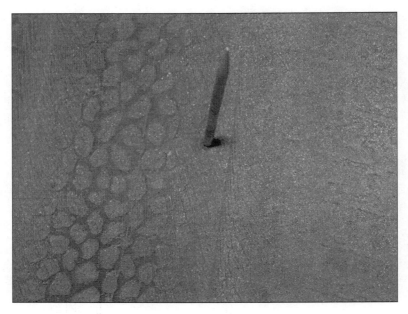

Fig. 4.6 Interior view of Fig. 4.5.

Identification of Causes and Contributors

Today's modern radial tires tend not to rust along the belt wire length. Although this possibility exists, rust formation today is primarily a somewhat local event that works at decreasing the integrity of the belt at and near the puncture site. The entry of salt into the wound can greatly degrade the belt structure in a much more rapid manner than water alone.

- **Damage to the belt system**—A small-diameter puncturing object (e.g., a small nail, a screw, or a piece of wood) may not cause many broken belt wires. However, broken belt wires are broken ends with rubber fracture and without the benefit of design parameters as seen at the belt edges to negate or slow the advance of the start of a separation. The belt ends continue to move as the tire rolls,[4-3] adding this effect to the others previously mentioned here. Larger-width damage (Fig. 4.1) that breaks multiple belt wires can greatly affect the starting point of a separation, even within the more inert center of the tire.

- **Improper repairs**—Improper repairs to a puncture by location, width, or type can cause conditions leading to a belt separation. See Ref. 1.8 in Chapter 1 for proper repair procedures.

Punctures that occur within or near a parabolic or belt edge separation can be considered *prima facie* evidence of the cause of the belt separation, and a puncture within 90 degrees of the separation very likely could be the cause. All pieces of evidence, including notes about and photographs of the tire, are needed to support this statement. To make the determination that a puncture caused the belt separation, all the items in this section should be considered, as well as the forensic expert's background and experience. Even if the puncture is patched and plugged correctly, you must look for the visual clues that determine if the puncturing event was the cause of the separation, including impacts and over-deflection. Real-world events can cause a separation to begin from a puncture, even if the puncture is 180 degrees from the belt separation.

For example, the following may be evident:

- Small punctures that cause small leaks may go unnoticed for some time, creating chronic over-deflection.

- If the vehicle owner is conscientious about adding air to the tire but does not spot the puncturing object, there might be several cycles of air inflation and deflation before the puncturing object is removed and the tire is repaired.

[4-3] During "proper" use of the tire, the middle third of the belt system is somewhat inert to high strains; however, the outer third on each side will have more movement (i.e., strain), especially near the belt edges.

- If the owner is not conscientious about tire maintenance, the tire could have been allowed to run to 100 kPa (15 psi) or less before the puncture is found and repaired.

- When an owner discovers that a tire is completely flat and takes the vehicle to have the tire repaired and aired up again, that tire could have been damaged even if it was not run-flat. Check for plug-only repairs that can be made without removing the tire, thereby negating an inspection of the interior. Plug-only repairs are against Rubber Manufacturers Association (RMA) practice, as is non-inspection of the interior of the tire.

- Some punctures also contain road hazard impacts; therefore, the impact section also should be reviewed.

Buttress and upper sidewall punctures should always be investigated, in addition to tread punctures. Although few punctures occur in the upper sidewall or buttress, these areas can be overlooked. Because of its lesser stiffness in comparison to that of the tread, the sidewall can be hand-manipulated to a greater extent than the tread, potentially opening a puncturing hole that can be seen on the liner side.

Upon tread and belt detachment, punctures can leave a hole with a "fluff" of ply cord (Fig. 4.7). An area of this type can be caused by accident damage or a puncture. Generally, accident damage will involve cuts, tears, cracks, gouges, and damage trails that may go through the "puncture" being investigated. Experience is required to denote the nuances and the differences between the accident damage and the puncture, if only to explain why this is a puncture hole and not damage. The area being investigated (i.e., the cords and what liner may exist) should be physically reconstituted back to its original shape as closely as possible. A puncture hole tends to be somewhat round, with possible ragged or torn edges, and a liner hole should line up with this area. If the tread exists and is located (i.e., spotted) versus the casing, a hole in the casing from a puncture also must have a hole in the tread or the sidewall.

Due to the typically severe damage of a run-flat condition, run-flat-type punctures may be rounded or jagged and nearly any shape, and they can be difficult or impossible to locate.

4.1.1 Over-Deflection

See Section 4.2 (on over-deflection) of this chapter for complete evidence identification.

Identification of Causes and Contributors

Fig. 4.7 A puncture in the upper sidewall.

4.1.2 Intra-Carcass Pressurization

Identification of intra-carcass pressurization (ICP) may be done as follows:

- **A separation under the #1 belt**—(Fig. 4.8) There can be degraded rubber and small tears visible in the carcass cord rubber in the shoulder areas under the #1 belt. When the "lifted" #1 belt edge is picked up, the area underneath will not show good tears but can be a rather smooth surface or may have very short tears from degraded belt stock. At times, there will be tearing of the #1 belt off the casing for a small distance inward, due to the "ripping" of the #2 belt and tread off the #1 belt during a tread and belt detachment. This is not evidence of ICP.

- **The #1 belt is detached, and the casing cords are visible**—Usually, there are portions of degraded rubber in the shoulder wedge area and/or across the crown, or the casing cords are bare of rubber. In Fig. 4.9, the #1 belt can be seen in the lower left corner. The #1 belt has been torn off the casing just above this area, as can be seen by the #1 belt rubber tears remaining on the casing. (Some ICP is noted in this area.) Above the #1 belt rubber tear area is the

Fig 4.8 Lifted #1 belt off the casing.

Fig. 4.9 Intra-carcass pressurization (ICP) evidence across the casing at the crown.

Identification of Causes and Contributors

smooth area without rubber tears, indicating ICP. Although this area of ICP is quite evident, some ICP locations will not be this obvious. The location of ICP at a liner split, as seen in Fig. 4.10, is fairly typical and is brought about by a breach in the innerliner typically from a road hazard impact. (See Section 4.4 of this chapter for detail and references about the topic of impact.)

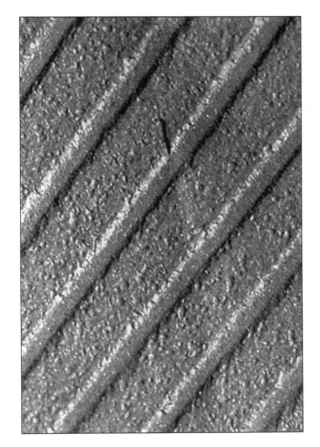

Fig. 4.10 A liner split less than 5 mm (0.2 in.) in length from a road hazard impact.

- **The #1 belt is adhered to the casing**—Here, the belt tears on the top surface of the #1 belt run in a circumferential direction (Fig. 4.11) rather than a basically radial to 45-degree angular direction. This rare kind of tearing may occur with physiologically damaged tires.

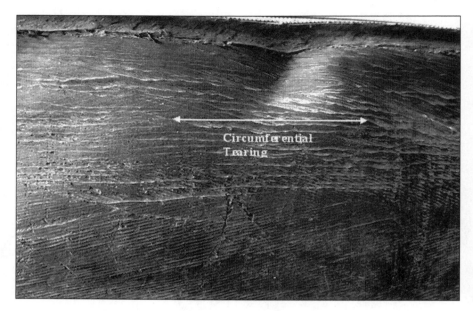

Fig. 4.11 Rubber tearing in a circumferential direction.

- **Oxidative appearance of the tire, including the belt compound**—Chain of custody and storage can affect this appearance. See Section 4.5 later in this chapter for a discussion of ozone deterioration.

- **A portion of the mid- or upper sidewall is detached**—This indicates the possibility that a bulge existed in this area (Fig. 4.12) prior to failure. The

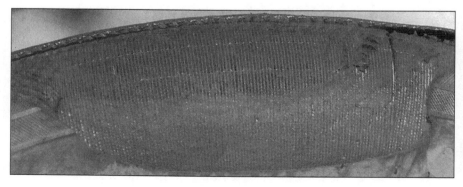

Fig. 4.12 The upper sidewall detached from the tire casing.

Identification of Causes and Contributors 51

bulge is formed by the sidewall becoming separated from the casing due to excessive air bypassing the liner. This area of detached sidewall and/or buttress typically will be torn in an oval or elliptical-appearing shape. Further tearing of the sidewall from the casing also can destroy portions of its original shape. Examination of the ply cords in this area may reveal the original shape by areas of slight polishing of ply cord rubber or continuous tears, such as stop/start marks as the bulge expands, or the lack of any marks in the original bulge area followed by tears in the ply cord rubber.

- **A portion of the upper sidewall and/or buttress is detached circumferentially around the tire**—Here, there are indications that the tearing of the sidewall off the casing cords was not as it should be in a tire without ICP. This is a judgment call, based on experience. Proper tearing of rubber off the ply cords typically reveals rough tears off the cords, as well as between the cords. Some examples of tears that are not as they should be are little visible rubber gauge remaining on the casing cords or the appearance of no multi-planar tearing.

Obviously, for ICP to have occurred, there must be a breach of the liner.[4-4] Because the liner runs from the inside of one bead toe[4-5] across the crown to the opposite side bead toe, liner breaches and the resulting ICP can come from any breach of the liner in any location on the interior of the tire, including bead toe tears.

Five possible reasons for ICP to occur are as follows (in no particular order):

- Liner splits from road hazard impacts (see Section 4.4 later in this chapter and Fig. 4.13)

- Mounting and/or demounting damage (see Section 4.6 later in this chapter)

- Liner cracking (including liner splice openings) (see Section 6.4)

- Punctures and/or improper repairs

- Advanced physiological damage

The identification of the cause(s) of the ICP should be noted, for instance, the liner breach (leaving out advanced oxidative damage). In some cases, this may

---

[4-4] One exception to this statement is physiologically damaged tires, where natural air migration has degraded the belt rubber.

[4-5] The liner ending at the bead toe is an approximation. Some tires have this ending around the entire bead into the wheel flange area, and some are short of the toe by some small distance.

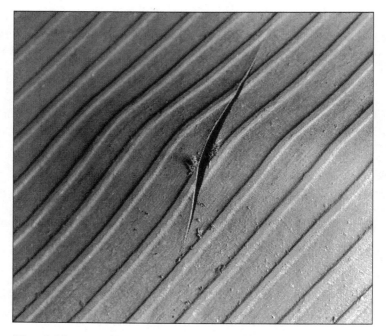

Fig. 4.13 A road hazard impact split of the liner.

not be possible, either because too much damage has occurred to the tire, and/or the breach simply cannot be identified. However, the clues that the liner breach existed as identified in this chapter will still be present.

Intra-carcass pressurization is a time-dependent event. Although a casing can be fully pressurized from the internal air pressure within three days of a liner breach, the oxidation of the belts takes time to occur.[4-6] In tires, there is always a pressure gradient from the internal tire pressure through the tire until the air exits the tire exterior surface. This pressure gradient exists throughout the inflated life of the tire and is the reason that tires naturally may lose from 3.5 to 14.0 kpa (1/2 to 2 psi) per month. When a breach of the liner has occurred, the internal air pressure having bypassed the liner will pressurize the ply cords that are adjacent to the belt package and especially the #1 belt. The oxygen in the air will begin to

---

[4-6] The amount of time will depend on the tire construction, the belt compound properties, usage, and environmental factors.

Identification of Causes and Contributors

degrade the belt rubber in a much more rapid manner than would have occurred without the breach. Over time, this oxidative attack degrades the belt compound properties, making the tire more susceptible to belt separation due to other types of damage (e.g., road hazards, over-deflection).[4-7]

4.1.3 Water and Salt Corrosion

In the past, water entering a belt cable through a puncture or penetration could run, by capillary action, the length of the belt cord, virtually rusting the wire from the inside out. In today's radial tires, the belt cord structures used and the methods of creeling the wire generally reduce the capability of the belt cables from rusting through their entire length. Thus, the rusting typically is a local event (Fig. 4.14), usually within several centimeters (inches) of the puncture or penetration. However, a visual rusting pattern even within several centimeters (inches) of the puncture can degrade the belt structure, which can work into a belt separation. Note that corrosion of the brass plating occurs before the rusting is visible.

Fig. 4.14 Rusted and broken belt cables at a puncture.

[4-7] Intra-carcass pressurization (ICP) alone eventually can cause a belt separation.

If the penetration opening is large enough and the penetrating object has been ejected, the remaining hole should be examined for rusting belt wire. Examination of a puncture or penetration wound can be accomplished with 4x to 7x magnifiers. With the top belt detached and supplied, this is a much easier environment in which to inspect the hole on both sides of the top belt, matching it to the #1 belt penetration or puncture location. If the detached belt is not supplied, then rust may be seen on the remaining belt or ply cords. When rust on the remaining belt or ply compound (Fig. 4.15) is found, then the hole probably allowed water into the belt structure. Be careful to examine the rust formation on both a macro basis and a micro basis. Storage conditions of the failed tire may have put it in contact with a metal object that was rusting. Objects that rust and have laid across or on top of the disabled tire typically leave a rust pattern that is not consistent with wire rusting (i.e., they leave an outline of their shape). If it is important to the case, the chain of custody, timing, and storage conditions should be investigated.

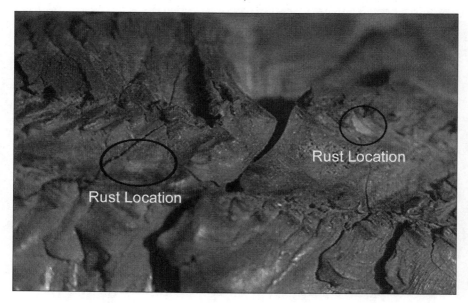

Fig. 4.15 Rust on a ply compound.

Determination of old rust versus new rust by color alone is difficult, requires experience, and is not completely accurate. It is better to judge the color or texture of the rust on various places on the tire and the degree of oxidation of the steel (with magnification). If the tread is only partially detached, x-rays of the attached

Identification of Causes and Contributors

(intact) portion can reveal wire oxidation (rusting) if it exists. Wire rusting in tread attached and therefore non-exposed areas of the disabled tire can eliminate the effect of storage contamination from other metal objects. Figure 4-16 indicates rust in a separated area that was noted on an x-ray and sectioned.

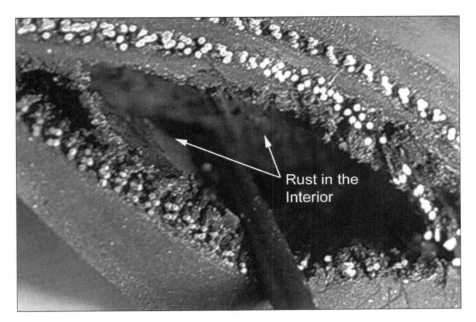

Fig. 4.16 Rust within a sectioned belt separation.

Salt deterioration almost always occurs in northern U.S. winter conditions. Salt is an extremely corrosive material to belt wires, and the entry of salt into the belt structure will accelerate the breakdown of the steel wires.

4.1.4 Breakage of Belt Wires

When belt wires break from a puncture or a penetration (Fig. 4.14), depending on the location of the broken or cut wires as well as the size and location of the puncture, the normal movement of the belts during the load and unload sequence of a running tire work the cut or broken individual wires at the puncture or penetration site. The closer the puncture or penetration is to the belt edge, the larger the flex of the broken belt wires. Excluding foreign materials entering an area of broken or cut wires, a puncture or penetration that breaks belt wires can cause a belt separation by the fracture of the belt compound (i.e., crack initiation).

Crack initiation from one broken belt cable in the middle third of the belt package would be unusual. However, the more cables that are cut or broken in this area, the more likely that crack initiation can start.

4.2 Over-Deflection

Over-deflection can be identified using various criteria.

4.2.1 Observation of the Compression Groove[4-8]

An expert can determine over-deflection by judging the amount of rubber displacement, scaling, luminosity, coloration, or cracking in the compression groove (CG). The CG area is identified in Fig. 4.17. (See Appendix C for additional details about the CG.)

When a loaded and inflated tire is deflected during contact with the pavement, there is a force moment in the lower sidewall and bead area, over the rim flange. This moment extends the sidewall out over the rim flange, putting pressure on the tire and wheel interface at the wheel flange radius. The rotation will cause a slight CG to form,[4-9] even when a tire is used in a proper manner. However, when this deflection goes out of the design standards, the over-deflection will cause deformation and compression of the rubber in this area to a much greater extent than normal.

As the over-deflection continues, this deformation and movement of the rubber deepens, and the CG expands and may include scaling, bluing, multiple rubber circumferential rings, and/or luminosity. The CG also may develop circumferential cracking from this constant bending and unbending of the rubber over the rim flange. As the over-deflection continues, the cracking in the rim flange bead area eventually can work through the rubber and fabric chafer, to and even through the ply cords. Extremes of this type of over-deflection can include cracking through both plies of a two-ply tire.

[4-8] Existence of the CG has been studied, noted, and/or documented in Refs. 4.1 through 4.4.

[4-9] A slight or slight-to-moderate CG will form from normal use. The forensic expert's background will determine the degree of normality because there currently is no bright line standard to use in judging the CG.

Identification of Causes and Contributors

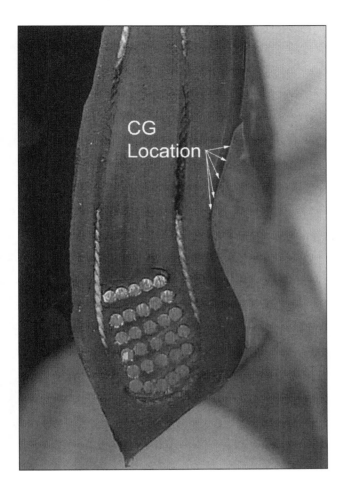

Fig. 4.17 A section view of new bead, indicating the general location of the CG.

The following are several notes on CGs:

- It is not unusual for the CG on one side of a tire to be different in appearance than the other side. In fact, it typically is the norm. This difference is affected by the road crown, vehicle-related conditions (e.g., camber), loading, and driving habits.

- The CG is a time-dependant event. For example, a punctured tire that suffers a fairly rapid deflation will not develop a characteristic CG from this severe but short-term over-deflection.

- The CG will not improve with time, even if the tire is no longer over-deflected for a period of its life.[4-10] Multiple CG rings may form, indicating multiple phases of over-deflection in the history of the tire.

- The CG is almost always in the same radius location as the tire balance weight marks. Because the formation of the CG and the tire balance weight marks are both time dependent, analyzing this area may yield a sequence of events in the life of the tire. (See Chapter 7 for a discussion of balance weight marks.)

- The CG will be different in nature, and its quickness to form will be dependent on the general construction of the tire, on a macro basis, as well as the aspect ratio, load, and pressure. Tires with stiff lower sidewall construction (i.e., some light truck types D and E load range tires) will need more over-deflection to cause a moderate CG than tires with less stiffness in the lower sidewalls. Under similar deflections, tires with short sidewalls (i.e., tending to be lower-aspect-ratio tires) will form a CG prior to taller sidewalls (i.e., having generally higher aspect ratios).

The variability of the CG with time, aspect ratio, and tire construction are a few reasons why there is no bright line standard for acceptability when judging the CG. However, most forensic experts with experience will arrive at similar conclusions.

4.2.2 Wrinkling of the Innerliner

The wrinkling of the innerliner in the upper sidewall (from non-static use) will be circumferential, with possibly multiple wrinkles of 360 degrees (Fig. 4.18). As the air pressure drops, wrinkling of the liner without any exterior circumferential sidewall abrasion will tend to occur on low-profile, short sidewall tires prior to its occurrence in tall sidewall tires. This is due to the short radius of curvature in the upper sidewall in low-aspect-ratio tires (e.g., P215/45R18) versus tall sidewall tires (e.g., P235/75R15). However, this appearance in the liner without any exterior evidence can occur in any tire.

4.2.3 Innerliner Color or Discoloration

Depending on the type of oil, various shades of blues, purples, and even deep greens can become visible[4-11] from the heat buildup that occurs when a tire is

---

[4-10] Tires do not heal themselves; therefore, all damage to a tire is cumulative throughout the lifetime of that tire.

[4-11] The heat buildup "bluing" discoloration of compounds can be found in most rubber areas of a tire, including the tread, sidewall, belts, and beads.

Identification of Causes and Contributors

Fig. 4.18 Wrinkling of the innerliner in the mid- to upper sidewall.

running on low air pressure [Ref. 4.1, p. 394] and with enough speed and load to build up excess heat that cannot be dissipated.

This bluing is not seen often in the innerliner of most modern tires due to less oil content in the liner versus that of older tires. However, if it does become visible, photographs should be taken as soon as possible because this bluing in the liner can disappear in time.

4.2.4 Exterior Sidewall Contact with the Road Surface

Contact of the upper sidewall (exterior) with the road surface will make a circumferential 360-degree, similar-radius abrasion ring on the exterior of the tire (Fig. 4.19). The 360 degrees can be an important time-dependent event because

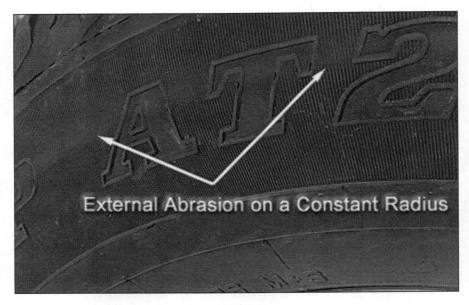

Fig. 4.19 An abrasion in the upper sidewall.

slightly inflated and still mounted tires make this ring, whereas a ring of less than 360 degrees may have had an air loss first and then the ring formed.

This event is almost always a low-inflation event; however, there is a load component to it. Thus, depending on the load, this abrasion line can appear on the exterior around 15 psi, but more likely it will be less than 10 psi. Round-shouldered tires (large shoulder radii and usually passenger car tires) will allow sidewall contact with the road surface at higher pressures[4-12] than square-shouldered or small-shouldered-radii tires (usually light truck-type tires).

4.2.5 Observation of the Tread Shoulders

If both shoulders are rounded from their original shapes, the tire may have been chronically over-deflected [Ref. 4.5, pp. 30–31].

[4-12] This discussion is still dealing with tires at 10 psi or less (depending on load).

Identification of Causes and Contributors

Large-shouldered-radii tires (generally passenger car tires) will show this more rapidly than square-shouldered tires (generally light truck tires). Both shoulders will seldom be rounded to the same degree. Driving habits, road crowns, and vehicle alignments can affect the degree of rounding from one shoulder to the other. This observation of the rounding of the shoulders should be made independently of any alignment conditions present in the tire because over-deflection and an alignment condition can be present at the same time. In addition, normal wear of the front positions on a front-drive vehicle naturally will wear the shoulders more than the center of a tire. Therefore, these other effects must be taken into account when determining that the rounding of the shoulders was due to over-deflection.

4.2.6 Wheel Weight Clip Mark Depth

The wheel weight clips used to balance a tire leave impressions in the rubber just below the bead centering ring (Fig. 4.20). The deeper the clip marks (i.e., indentations into the rubber), the more over-deflected the tire was during its history [Ref. 4.1, p. 394]. The CG and the balance weight mark are located at nearly the same radius within the tire and physically will overlap each other in most cases.

Fig. 4.20 A deep balance weight mark.

The CG should be used as the primary evidence of over-deflection, with the balance weight marks as the secondary evidence. Making the balance weight mark secondary to the CG does not mean that mark should be ignored. Although good correlation usually exists regarding indications of over-deflection between the depth of the clip mark and the CG, the clip placement over the flange may cause some variability in the depth of the rubber.

Because there is no universal standard for the depth of balance weight marks, a simple "light, medium, and deep" connotation can be used to describe the depth of the clip marks in a tire.

Balance weight marks will not disappear over time unless the weight is removed. Then over time, the CG may override all or part of the balance weight mark. The converse also is true, as the balance weight mark likewise can form over the CG (Fig. 4.21). Deciphering which mark came first is a matter of determining which feature overrides the other (i.e., was last in place).

Fig. 4.21 Light balance weight marks that appear post-creation of the CG.

4.2.7 Shifting or Chattering of the Balance Weight Mark

A tire slipping on the wheel can be indicated by shifting or chattering[4-13] balance weight marks. A shifting is a movement of the weight and reformation of the balance weight mark. Chattering is considered multiple movements. This shifting and chattering (Fig. 4.22) can occur with low inflation pressure during the life of the tire, high braking or accelerating forces applied soon after mounting the tire to the wheel, or if the bead seat is disturbed temporarily due to impact. Braking forces that can move a tire on the wheel after the mounting soap or lube is dry normally will leave a brake skid mark in the tread.

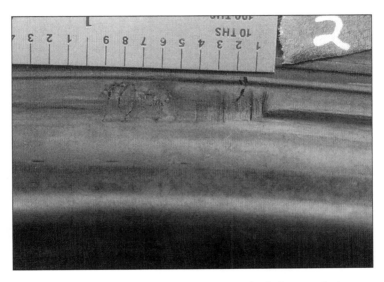

Fig. 4.22 Balance weight movement, with shifting and chattering visible.

However, over time, these initial braking or accelerating forces that may have left evidence on the tread as striations or abrasions may be long worn off.

In some general form, the balance weight marks will remain. Rebalancing the tire can give a similar appearance, on a macro scale, to a shift of the wheel weight;

4-13 Note that the tire slip leaving shifting or chattering marks is the time-dependent event that leaves solid indentations into the tire, not the sharply defined marks of the accident sequence.

however, rebalancing is a placement of the weight on a rim flange and not a shifting of the weight. Thus, there will be a different appearance of the rubber indentations at the edges of the balance weight mark, albeit sometimes slight.

4.2.8 Interior (Liner Side) of the Bead Toe—Creased or Cracked

In passenger car and light truck tires with fitments on 5-degree taper wheels,[4-14] deformations in the bead area due to the rocking motion of the bead on the rim (i.e., over-deflection) can form a crease and/or a crack on the inside of the bead toe region (Figs. 4.23 and 4.24).

Fig. 4.23 Creasing on the liner side of the bead toe.

[4-14] To determine the tire and rim fitments and the rim seat profiles, see the latest *Yearbook* of the Tire and Rim Association (TRA) (Ref. 1.6 in Chapter 1).

Identification of Causes and Contributors

Fig. 4.24 Cracking of the liner side of the bead toe.

During over-deflection, this rocking in light truck tires with a wide bead face[4-15] (typically, LRD [load range D] and LRE [load range E] tires) can cause a small reverse bead angle to form. (See Appendix B for a definition of reverse bead angle.) The reverse bead angle typically is a medium/heavy truck tire phenomenon, with 15-degree bead seats.

This creasing and/or cracking as shown in Figs. 4.23 and 4.24, respectively, generally will not occur on tires with 15-degree wheel bead seat tapers. On tires that are placed on wheels with 15-degree bead seats, the rocking will reflect itself on inspection as a dual angle present in the bead face of the tire. The original wheel taper angle of 15 degrees, starting at the bead heel of the tire, will be present, with a second angle starting at the mid-bead face. The second angle will be smaller than 15 degrees. The smaller the second angle during the original tread life,[4-16] the greater the over-deflection. (Some secondary taper angles will go to 0 degrees or to negative angles.)

4-15 These generally will be commercial-type light truck tires and medium/heavy truck tires.

4-16 Medium/heavy truck tires used in a proper manner will develop this dual bead angle as a natural occurrence following a very long life of the casing (i.e., one or more recaps). Formation of a significant reverse bead angle during the first life is over-deflection. Experience in medium/heavy truck tires is a necessity to make the correct judgments.

4.2.9 Bead Face Abrasion or a Circumferential Line in the Bead Face

In tires with wide bead faces and stiff lower sidewalls (i.e., light or heavy truck tires), another indicator of over-deflection is an abrasion located approximately in the middle of the bead face, circumferentially around one or both beads. This can appear as a line or a crease in rubber chafer tires, or at times, the compound of the fabric chafer will abrade down to the chafer cords or ply cords. In Fig. 4.25, this can be seen as a wide area of abrasion. The compound in this particular tire in the bead face actually has become brittle from the heat exposure.

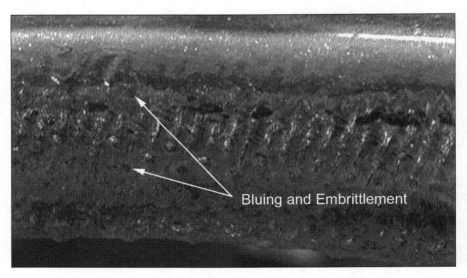

Fig. 4.25 Bead face abrasion, with fabric visible and bluing from heat.

4.2.10 Creasing of the Base Radii of the Tread Grooves

Creasing of the base radii of the tread grooves (Fig. 4.26) can occur with over-deflection caused by under-inflation. However, it also can occur if the tire is overloaded but the internal air pressure is at vehicle placard or higher (primarily LRD and LRE load range and medium/heavy truck tires). The overload and the high pressures compress the tread lugs, causing creasing at the base.

Identification of Causes and Contributors

Fig. 4.26 Creasing at the groove base.

4.3 Penetrations

4.3.1 Tread Attached to the Casing

Penetrations are defined as anything that penetrates the tread, buttress, or sidewall from the exterior and may continue deeper into the tire but does not reach the air chamber. A penetration may leave a mark, cut, tear, hole, or so forth in the tread (Fig. 4.27) or buttress surface. Penetrations range from cuts, tears, snags, and so forth to hard objects that have penetrated the tread, may or may not still be enveloped in the tread, or may have been ejected from the tire. Apart from the technical meaning of a penetration, this term in tire forensics generally is not used broadly for tears, snags, or gouges but rather for discreet holes (Figs. 4.27 and 4.28) or objects that are still stuck in the tread. The identification of penetrations that reach the belt package—by location, size, and depth—is necessary so that a full picture of the environmental use of the tire can be put together. Those

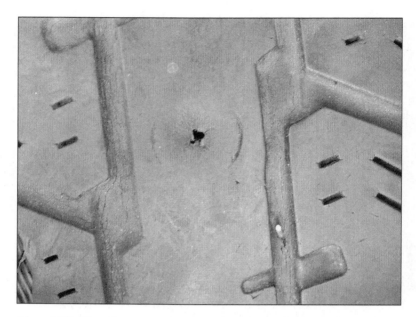

Fig. 4.27 Tread penetration into the belt package, with a circular mark surrounding the hole.

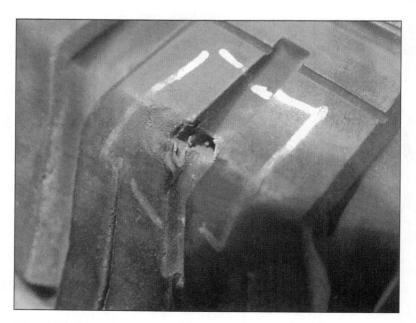

Fig. 4.28 Penetration into the belt package in the tread shoulder.

penetrations that damage the tread may be noted if they are significant to the usage environment or to an area of interest.

Penetrations are probed, if allowed, in the same manner as a puncture. However, the goal in this case is to feel for the belt package. Be particularly careful in probing penetrations. Care is always needed when probing an opening in the tread. One does not want to accidentally puncture a tire when the penetrations go through the tread, both belts, and the casing, leaving only the soft rubber liner not penetrated. Penetrations that are found only in the tread compound, depending on the size, depth, amount, and type, can tell you something about the usage environment of the tire (Fig. 4.29). The greater the number, type, and size of the penetrations, the more likely that a puncture or road hazard impact might exist, even if the penetrations are found only in the rubber.

Fig. 4.29 Multiple penetrations within the tread rubber.

Penetrations that reach the belt package are detrimental to the tire in the short or long term. By opening the penetration area for visual inspection or feeling the belt package with a dull tool, a judgment can be made whether the penetration is through one or both belts. The deeper and wider the penetration into the belt package, the worse the wound is, allowing foreign material into the area and potentially rusting the belts.

Generalized cutting and chipping of the tread is a special case of penetrations, although the terms cutting and chipping usually are used separately to denote the type of environment in which the tire was used (see Section 4.9 later in this chapter).

4.3.2 Tread and Top Belt Detached

If the top belt and tread are available for inspection, the detached piece(s) first must be matched to the existing casing and then the bottom side of the top belt inspected for penetrations that have breached that belt.

Penetrations in the top belt will be visible and apparent during the inspection as a hole with possibly cut or broken cables (Fig. 4.30). Rust also may be visible. Sometimes a cut or broken wire cable will be evident as a track (Fig. 4.31). A track is left on a belt when a cable in either belt has been cut or broken previously, and then the tread and top belt detach, pulling the cord out from one or the other side of the cut or broken wires.

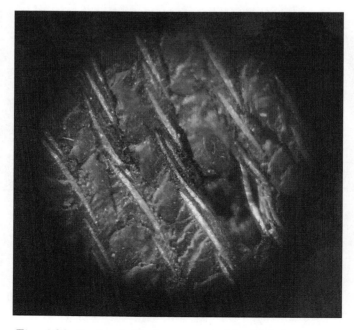

Fig. 4.30 Broken belt wires on top of the #1 belt.

Identification of Causes and Contributors

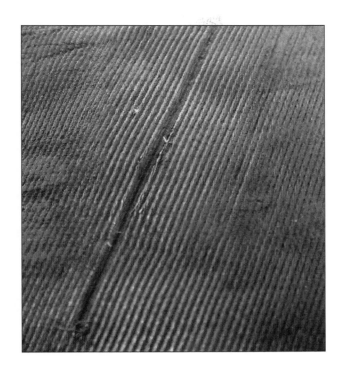

Fig. 4.31 Penetration with a missing wire (a track).

During the tread and belt detachment, a broken belt cable will remain with one belt or the other. This leaves an empty space where the cable should be on one belt, and an "extra" cable on the other belt.

If the detached tread and top belt are not present for inspection, it is more difficult to find penetrations that went through the top belt because the only ones you will find are those that reached through the top of the bottom belt or that left a mark on the surface.

Penetrations through the top belt that have not penetrated the bottom belt can be identified as follows:

- **A track from the top belt is left on the bottom belt.**

- **Dirt, grime, and/or rust is present in a localized area that has some defined shape to it, with or without a small gouge or mark into the compound of the belt**—(Fig. 4.32) For instance, if a 12.5-cm (1/2-in.) long straight cut through the tread and top belt stops at the bottom belt, grime will be deposited over time onto the bottom belt compound in the shape of a straight cut

Fig. 4.32 A gouge mark at a penetration.

on top of it. When the tread and the #2 belt detach, this mark remains on the #1 belt.

- **Rust is visible on the rubber**—Broken top belt wires can rust when water enters the opening, depositing a rust mark on the compound (Fig. 4.33) of the bottom belt. This is not an insignificant find because rubber does not rust, and a rust deposit on rubber indicates the probability that the wires were rusting. Careful inspection of the shape of the marks and/or the location of the rust is required to make the determination that the rust is from the belt package rather than from storage conditions post-accident. (See Section 4.1.3 for further information about corrosion.)

A few ways to separate rust from storage versus belt rusting are as follows:

- A metal object that would lie against a detached tread belt surface generally will leave its own impression on the surface, even if the rust gets into the crevices of the rubber.

- Rust from the belts will leave its deposits within crevices where the belt cords were located.

Identification of Causes and Contributors

Fig. 4.33 Rust on rubber in the belt region.

- If part of the tread or belt remains attached to the tire, an x-ray of that area can find rusting belt wire if the rusting is in an advanced stage.

- If an area exists where the tread and all the belts are detached, the casing cord can "fluff" up (Fig. 4.34) in a very localized area. This occurs when the penetrating object reaches the ply cords, and then on belt and tread or sidewall detachment, the broken fabric ends are pulled in an outward direction, revealing the penetration.

The expert must rely on experience and visual acuity to analyze the overall area with cord fluff and to pick out what could be a penetration to the post-tread detachment damage to the cords.

4.4 Impacts

Road hazard tire failures can vary from the immediate to months and many miles after the initial impact [Ref. 4.6, pp. 4–5]. A road hazard impact that breaks significant ply cords and belt wires generally will be a near-term failure. Figure 4.35 shows a tread compound fracture from an impact that led to broken ply cords

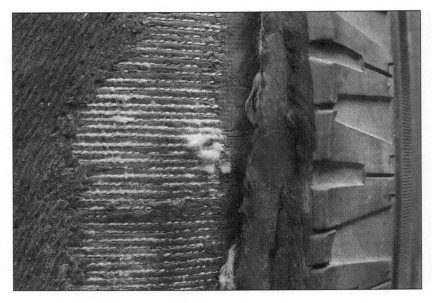

Fig. 4.34 Cord fluff at a penetration.

Fig. 4.35 An impact fracture of tread compound.

Identification of Causes and Contributors

(Fig. 4.36) and a run-flat condition. However, not all impacts result in broken ply cords, broken belts, or tread fractures.

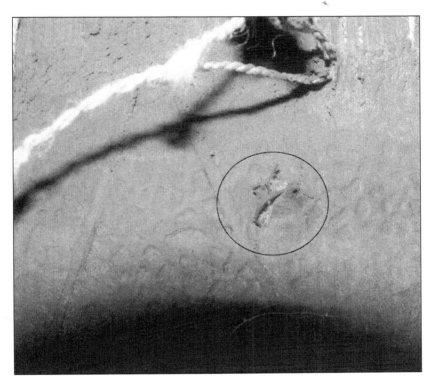

Fig. 4.36 Broken ply cords, viewed from the liner side of Fig. 4.35.

Impacts that fracture the belt compound in a local area or that cause small radial innerliner splits [Ref. 4.7, p. 6] also can cause belt separations over time. The amount of time depends on many factors, including environmental factors (e.g., roads, driving habits, speed, vehicle suspension, ambient temperature) and the current condition of the tire (e.g. load, air pressure, tire integrity).

Two examples of impact timing to failure that I have viewed are as follows:

- Traveling on I-5S in some fairly dense Los Angeles traffic, the front tire on an S-10 pickup truck, approximately four car lengths in front and one lane to the right, kicked up what appeared to be a metal plate. The left front tire initiated the plate movement, and the left rear tire caught the plate on an angle to the road surface. The resulting impact failure was nearly immediate, and the S-10 pulled safely off the road.

- Near sunset, a shadow on a tire caught my eye, with a pencil-appearing radial bulge in the mid- and upper sidewall. It quickly was discovered that the owner remembered running up a curb several days prior. After demounting the tire from the wheel, it was noted that the pencil line was the result of broken ply cords due to a rim pinch with the curb. Investigating the wheel, some minor damage to the rim flange was noted, which by itself would not attract any attention. Although the tire had not yet failed, this pencil bulge eventually would result in an immediate air loss at some time in the future.

4.4.1 Identification of Road Hazard Impacts

4.4.1.1 Fractured Tread or Sidewall Compound—No Immediate Failure

When tread rubber impacts an object or hole in the road with sufficient force, it can fracture [Ref. 4.8, pp. 45–58] in spidery lines off the impact point. These fracture lines are distinctively different than the ozone deterioration lines that form over time in the tread grooves or lugs and the sidewall. There may or may not be visible damage to the structure of the belts [Ref. 4.9]. If the tire runs for some time after the impact, ozone deterioration will tend to enhance the appearance of the fracture (Fig. 4.37). Ozone deterioration around the fracture point takes time to develop, and its presence in the impact strike area can place the fracture prior to the accident sequence.

Ozone deterioration lines will tend to follow the groove and base radii of the lug, groove, and sidewall letter patterns. A shift away from the normal weathering lines at the base of a groove can be an indication of an anomaly such as an impact (Fig. 4.38).

Objects that are caught in the tread and work into the grooves of the tread also can distort the ozone deterioration lines at the base of the groove. The need is to distinguish that the distortion in the grooves is from the impact fracture and not the time-dependent pressure of an object stuck in the tread. Typically, objects trapped within the tread grooves generally are stones and, while working their way down to the groove base, will leave impressions on the side angles of the lugs, without fracture lines.

A bias tire tread impact failure leaves a fairly characteristic "X" mark [Ref. 4.10, p. 181] in the ply impact area. This "X" appearance is due to the failure following

Fig. 4.37 An old upper sidewall impact exposed by ozone deterioration lines.

Fig. 4.38 Tread fracture within a groove.

the "X" pattern of the cross plies. This type of "X" break was and is well known to many who analyzed tires in the "old bias tire days." A radial tire with radial ply lines does not follow the "X" break concept except at those times where the "X" break pattern approximately follows the belt cord angles (Fig. 4.39) for a short distance.

Fig. 4.39 Impact damage of the tread with a broken belt.

Generally, the tread belt detachment appearance from an impact in a radial tire is one that does not follow the belt angle line of a "clean"[4-17] belt separation from tread edge to tread edge. At times, one belt breaks, and the other one splits open. It is not uncommon to find the tread ends on both sides of an impact-caused separation to be "ragged" in appearance. Accident-caused damage in the tread area tends to be slices, cuts, tears, and abrasions. Tread breakup and ragged tread ends due to impact with the underside of the vehicle must be investigated and ruled out.

4-17 A "clean" belt separation refers to a tread and belt delamination where the tread and top belt detach from the bottom belt, and the initial separation flap generally follows the angle of the belt pattern. A ragged belt separation is one in which there is rupturing occurring in an area. Although a belt separation may be underneath, the beginnings or endings of the tread detachment generally will not follow the belt angles.

Identification of Causes and Contributors

Impact-type damage requires a "rock and a hard place." When a tire is inflated, the interior air pressure provides the hard place, while the object is the "rock." When a tire is deflated, there is no hard place to cause the impact-type fractures unless the tire, during the accident sequence, happens to be caught between the road surface and another object such as the wheel. When considering a disabled and deflated tire that most likely is unbuttoned (debeaded) and is "flopping" around during the accident sequence, damage trails prior to and/or after an impact fracture should be investigated.

A damage trail leading to and/or after an impact fracture most likely will be formed from an event that occurred prior to the accident. If the tire becomes unbuttoned from the rim, sporadic and random damage tends to be the norm during the accident; however, if possible, one should always look for time-dependent events to trace pre-accident damage versus damage that occurred during the accident sequence. Identifying the impact fracture as occurring prior to the accident sequence can be done visually in a few ways.

Fractures on the tread surface area obviously will continue to be worn as the tire rotates and can be obliterated completely over time prior to tire failure [Ref. 4.7, p. 5]. In Fig. 4.40, the tread surface has continued to wear, while the sides of

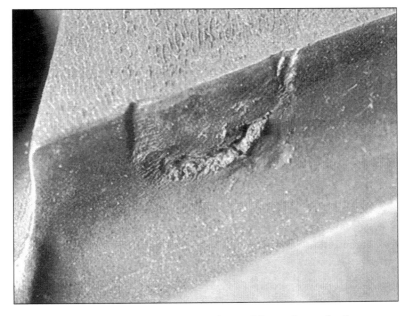

Fig. 4.40 A worn tread surface with tread lug side angle damage.

the lugs maintain the original damage. In Fig. 4.41, the puncture area still has sharp defined lines while the tread continues to wear. The degree of wear and the appearance of the tread surface versus the sides of the lugs where wear is not occurring can be a determining factor in the sequence of events prior to the accident.

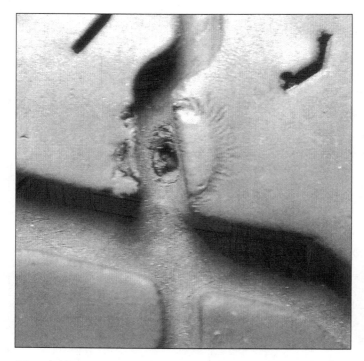

Fig. 4.41 A worn tread surface with a sharply defined puncture area.

An impact just prior to or during the accident sequence will show sharp edges and no wear on the tread surface. Cycles on the road tend to dull the rubber edges, and ozone deterioration eventually can move through these areas. In Fig. 4.41, the tread surface cracks are rounded and less defined than those in the grooves; however, the degree of rounding on the surface is not that great. Thus, the road hazard impact that caused the tread fracture did not occur a substantial amount of miles prior to the failure.

4.4.1.2 Fractured Belt Compound—No Immediate Failure

Fractured belt compound from rapid impacts can lead to belt separation. Although it seems logical that rubber with its relatively high elongation should by nature fracture much later and at higher strain than the steel belt cables, this is not necessarily the case. With rubber fracture, it is not the final elongation that determines the fracture point; it is the strain rate. Strain rate is the amount of deformation of a body, dependent on the force applied and the shape of the object applying that force, per element of time. Current ASTM testing for elongation (D412-06a) stretches a rubber component at approximately 5 cm (2 in.) per minute. An impact on a tire moving at 27 m/s (88 fps–60 mph) is orders of magnitude higher in strain rate. Belt compound testing of this type is not found in the general public domain. Bekar *et al.* (University of Akron) published an article [Ref. 4.8] describing this kind of strain rate for rubber fractures of styrene butadiene rubber–synthetic rubber (SBR) type compounds. Although belt compound is made from natural rubber (NR), belt stock fracture does occur [Ref. 4.11, pp. 252–275] without belt cord breakage. Standard Testing Laboratories (STL) found this to be the case empirically with its SAE J1981 test after revising striker types to change the impact type [Ref. 4.9, pp. 7–8]. In making this change, one of the impacts noted was that a belt cord could become loose (rubber fracture) within the rubber matrix without belt wire breakage. Impacts that occur toward the middle of the tread with sufficient force to fracture belt compound also may break belt cables and/or ply cords. However, as seen in the STL test, this is not always the case.

Road hazard impact belt compound fracture can become a crack initiation point, especially in the highly strained belt edge areas. Depending on the dynamics of the impact and the continued environmental and tire issues (e.g., load, speed, pressure, vehicle), this area can become the site of crack propagation and growth.

4.4.2 Identification of Belt Separation Due to Impact

Although crack initiation and propagation are similar to many kinds of separation initiation causes, the identification of belt separation due to impact can be any combination of the following:

- **The area of belt separation is localized**—The belt separation is limited to an area of the tire, and belt edge separation (BES) is at a minimum, or causation for a larger area of BES can be determined and another major separation does not exist in the tire.[4-18] Road hazard impacts can occur with other belt

---

[4-18] On rare occasions, there may be two major separations in a tire.

separation causes and contributors that may seem to cloud the "localized separation" evidence. For instance, if over-deflection exists and could be considered causative to the failure, an analysis should be made considering whether the over-deflection or the impact was primary by breaking out further evidence and, if possible, the timing of each. Sometimes this is very difficult to do. For example, over-deflection can degrade the properties of a belt compound to the point that an impact of much smaller force than would be needed without over-deflection can start a belt separation.

- **Tread or upper sidewall or lower sidewall damage from just before to just after the separated area**—These markings can be gouges, tears, cuts, scuffing, abrasions, and/or rubber fractures. All of these markings should be interconnected to some degree by angle[4-19] and/or location. Under small magnification, these markings should be judged as coming from a direction of the tread to the rim. (See Section 4.8 on snags, gouges, cuts, tears, and abrasions.) Markings that leave a trail from the rim upward to the tread generally are accident damage, mostly from the rim.[4-20]

  Although punctures can cause belt separation 180 degrees away from the separation, an impact leading to a belt separation typically occurs in the general proximity of the separation. If the outboard side of the tire and the side of the vehicle where it was located can be determined, it is possible with the belt separation parabola to determine the likely areas for the impact to occur. (Tires that transfer from vehicle to vehicle or are flipped DOT side inboard and outboard make this difficult to determine, as do certain rotation patterns.)

  "Likely areas of impact" are stressed because while the rolling tire usually will cause the separation point to grow opposite the rolling rotation, this is not always the case. Impact points at the middle of the parabola or at the back edge from the direction of rolling do occur, although they are less common.

- **A small circumferential line in the bead face**—This line can be a mark or a compression-type tear. A compression tear is a tear that is formed from compressing rubber quickly and tightly. This is unlike a tear we see as a tensile tear of the rubber.

- **A mark or deformation line on the innerliner**—The mark generally is circumferential, linear in nature, and located in the upper sidewall or tread shoulder areas of the innerliner (Fig. 4.42). These marks are caused by the

---

4-19 Remember that the tire is rolling upon impact; thus, at times, the impact point that causes the crack initiation can be at a somewhat different location than the damage point.

4-20 An exception to this is off-road use.

Identification of Causes and Contributors 83

Fig. 4.42 A linear mark on the liner.

sidewall and tread portion of the liner deflecting from impact and pinching the rubber together or coming in contact with the rim. The liner marks also can be within the tread shoulders of the tire.

In contrast to a circumferential linear mark, Fig. 4.43 shows a circular mark on the liner. During inspection and matching of the tire to the wheel to their pre-accident location, it was visually apparent that this mark on the liner left a corresponding imprint on the wheel well.

- **A generally oval area in the upper sidewall, with the sidewall compound detached from the casing**—The oval area can appear as a bulge in an inflated tire. Figure 4.12 indicates the oval shape of a prior bulge located at a sidewall split. It is not unusual to see this combination of radial casing split with detached sidewall compound (Fig. 4.44) at a belt separation. This combination points to an impact that did not fail the tire until sometime after the impact event.

- **A bending of the rim flange in an outward (outboard) direction from radial**—An impact can deform the rim flange outward, while leaving little or no evidence on the tire. Therefore, inspection of the wheel and match locating

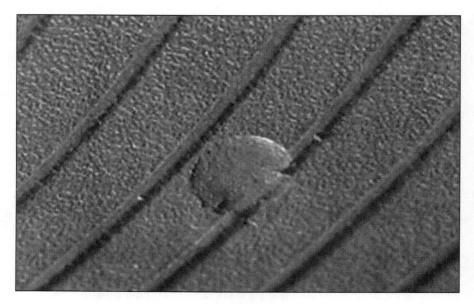

Fig. 4.43 A liner mark from impact with the wheel well.

Fig. 4.44 The upper sidewall detached.

Identification of Causes and Contributors 85

the tire to the wheel can be an important part of the analysis, if impact failure is being considered as the failure cause. In Fig. 4.45, note that in addition to the outward deflection of the rim, there are two distinctly different circumferential abraded flange areas, as noted by the black paint removal. This indicates two distinctly different over-deflection events.

Fig. 4.45 A bent rim flange.

- **Radial casing split(s) in the shoulder area of the tire**—A radial casing split due to an impact can be the location of what was a very small split in the innerliner. This small radial liner split is a breach of the innerliner along the ply cord path and begins the process of intra-carcass pressurization (ICP) [Ref. 4.6]. Because liner breaches and/or belt compound fracture tend to cause belt separations between the casing and the #1 belt and between the #1 and #2 belts, typically all of the belts and tread are detached in the area of this impact. The small liner splits typically become much larger radial casing splits during tread and belt detachment or during the accident sequence.

    Determining whether the casing split occurred prior to and not during or after an accident can be made as follows:

    - **Check the liner**—If substantial accident damage does not exist, the liner may have a "worked" appearance at the radial breach, as opposed to a new split appearance. This worked appearance comes from the opening and closing of the ply cords as the tire rotates in and out of the footprint. As an example, open your hand fully, and move your fingers in a motion to spread them apart and then bring them together. This is the motion the plies make through the footprint. The innerliner, being attached to the plies, makes this same working motion. In a tire that has sustained a tread and belt detachment and air loss with accident damage, this type of working of the innerliner may not be found because the accident sequence can cause damage at the edges of the radial split, mimicking this worked appearance.

    - **Check the tire for road rash**—(See Section 2.1.3 of Chapter 2.) A tire with classic road rash on the #1 belt surface was inflated after the tread and belt detachment. It is unlikely that an existing casing split (prior to the tread and belt detachment) would not split open further upon the detachment, causing a sudden air loss.

    The evidence of ICP and the lack of any other breach of the innerliner can be attributable only to an existing radial casing split. Although this seems as if eliminating other things yields this, it is intuitive and correct as noted here:

    - **Liner splits from road hazard impacts follow the ply lines**—When a tire loses its belts, this pre-existing split can split wide open because it is already an existing weak spot.

    - **By definition, ICP can occur only with a breach of the liner or substantial belt compound oxidation**—If no other reasons are found during an inspection (e.g., liner cracks, open liner splice, bead tears) for a breach

Identification of Causes and Contributors

of the liner, by default the radial split is the cause. (See Section 4.1.2 on ICP for further information.)

- **Detached upper sidewall compound (Figs. 4.12 and 4.44).**

- **Broken ply cords, belt wires, or loose belt cables within the belt compound**—As noted by Bolden [Ref. 4.9], impacts can cause a fracture surface around the individual belt cords, causing a looseness of the cables within the rubber matrix. In experiments, there was no damage to the exterior or interior of the tire from the impact. The tread was stripped off at the area of the impact, revealing the loose cables. By varying the striker head types, orientations, force, striker lengths, and impact location on the tire, the STL articles in Refs. 4.6, 4.7, and 4.9 indicate a broad spectrum of impact damage that can occur, leading to tire failures immediately upon impact or months and many miles later.

  Impacts that break belt wires or cables tend to break the belt wires in tension. However, some of the belt cables can be cut from the object being hit, or the impact has loosened the cables and there can be flex breaks noticeable. These types of wire fractures all can be identified with a 15x magnifier and inspection of the severed belt cord endings. Steel belt filaments that have been cut by an object will show a sliced appearance generally on an angle.

  Steel wires broken in tension indicate the traditional wire "cup and cone" tensile break. Flex breaks show a jagged break, whereas cut wires will have a smoother appearance (Figs. 2.16 through 2.18 in Chapter 2). If a tire remains inflated after a tread belt detachment and continues to run on the road, any exposed belt wire endings can become rounded from being ground into the road surface. Then, determination of the type of break may be impossible. The amount of rounding depends on the location of the ends across the tread surface, the amount of air in the tire, and any protection the ends receive from any remaining tread compound.

  Unless there is a breach of the innerliner with a direct path to the tire surface, broken ply cords and belt wires do not automatically result in a sudden air loss (Ref. 4.6).

- **Sharpness of the fracture area within the belt structure**—As noted by Daws, impacts tend to give a sharp fracture surface (Chapter 1, Ref. 1.2, p. 20) at the belt compound level. If the tire does not fail shortly after this fracture surface is formed, the sharpness of the fracture surface will become rounded as polishing of the surface sets in over time.

Items that increase the likelihood of sustaining a road hazard impact disablement are as follows:

- Increased air pressure [Ref. 4.6]
- Decreasing tread depth[4-21]
- Low-aspect-ratio tires with small-height sidewalls[4-22]
- Stiff belt packages (e.g., nylon overlay in addition to steel belts)

4.5 Ozone Deterioration

Ozone (O_3) is a gas that generally is associated with smog. However, ozone exists whether or not the air is smoggy. In addition to naturally occurring ozone, there are man-made sources as well, such as electric motors and welders and most transportation. Whether man-made or naturally occurring, ozone will deteriorate rubber on a molecular level. Tire rubber compounds are made of polymer chains, and ozone contact with the surface of some of these compounds eventually will "cut" the chains (i.e., chain scission), leading to crack formation (Figs. 4.46 and 4.47). The deterioration of tread compounds occurs in the same manner.

Fig. 4.46 Ozone deterioration in the upper sidewall/buttress.

[4-21] STL experience with its SAE J1981 test suggests that lower tread depths allow for a smaller impact force to cause tire damage.

[4-22] From the standpoint of the tire, there is less ability to absorb an impact in small-section-height tires.

Identification of Causes and Contributors

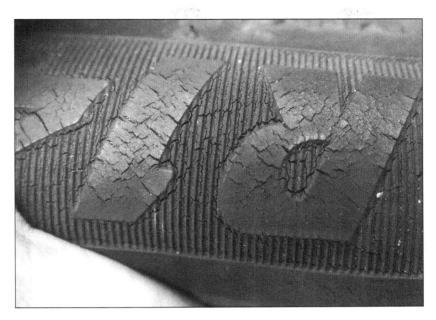

Fig. 4.47 Ozone deterioration in the mid-sidewall.

Ozone deterioration, or "weather checking" in layman's terms,[4-23] is the cracking of the rubber from the surface inward toward the structural parts of the tire.

Tire manufacturers incorporate anti-degradant materials into the various components of a tire. These materials delay the onset of and reduce the ozone deterioration, but they do not prevent it [Ref. 4.12, p. 1]. There are various types of anti-degradants,[4-24] and each does a specific job in preventing deterioration [Ref. 4.13, p. 1], whether the tire is standing still or rolling on the road.

Ozone deterioration is used as help and guidance when considering the environmental use of the tire, physiological aging beyond the chronological age of the tire, or in determining sequential events in the life of the tire. These determinations can be made due to several characteristics of rubber, which, when combined with ozone, lead to certain conclusions. These characteristics are as follows:

4-23 The other layman's term for ozone deterioration is "dry rot." This connotation, taken from the lumber industry, is not accurate in any sense of the term concerning tires.

4-24 Anti-degradants (e.g., anti-oxidants, anti-ozonants) can be found by name in Ref. 4.14.

- **Rubber will crack faster and deeper when it is strained**[4-25] **than when it is not strained**—For instance, a thin slab of rubber that is bent in a "C" position and exposed to ozone will form cracks at the fold point of the "C" faster and deeper than near the middle or ends of the "C." This fact helps identify areas of a failed tire that had been under a high strain.

  For example:
  - Lower sidewall cracking in a circumferential manner (Fig. 4.48) or cracking just below the buttress design in a circumferential manner can indicate that the tire has been run over-deflected if the degree[4-26] of cracking is more than the mid- or upper part of the sidewall.

Fig. 4.48 Lower sidewall cracking in a circumferential manner.

[4-25] In ASTM D-1149, the testing of rubber for surface ozone deterioration requires that the component be strained.

[4-26] I use the Shell rating scale, which indicates the cracking of the rubber from a level of "0" as the worst to "10," which is basically a new tire. (See Appendix E for more information on the Shell rating scale.)

Identification of Causes and Contributors

- Creasing across the sidewall in a straight line, similar to a chord in a circle, can indicate that the tire had a storage problem in an uninflated state or in a near-zero-pressure state. The same is true for a tire where the creasing has developed into a crack.

- The weight on an uninflated tire causing the strain (crease) in the sidewall could be from a mounted wheel or some other weight on an unmounted tire.

- Improper tire storage, such as a tire stored flat with sufficient weight stacked on the tire to bend and bulge the tread/belts circumferentially, can cause cracking in the tread grooves.

- **Rubber will crack faster and deeper when it is exposed to the sun (over time)**—The side of the tire that was on the outboard side of the wheel for the greatest amount of time generally will have the most ozone deterioration, even if both sides of the tire have the same general aesthetic appearance.[4-27] Exposure to the sun also can yield a higher durometer of one sidewall more than the other (i.e., the more likely outboard side is the higher durometer). Note that in whitewall tires, the white compound is covered with a thin veneer rubber piece called the "cover strip." This component is made of ethylene propylene diene monomer (EPDM) rubber, which is non-staining (protects the white compound) and is highly resistant to ozone deterioration. The cover strip is wider in a radial direction than the whitewall compound by approximately 12 to 25 mm (1/2 to 1 in.) more or less per side (Fig. 4.49). The demarcation line of the top edge (closer to the tread) of the cover strip is seen easily in an older tire because the cover strip still remains relatively free from ozone deterioration, whereas the rubber above the sidewall contains cracking. To take a proper measurement of the sidewall hardness on the white sidewall side of the tire, the durometer must be placed above the top ending of the cover strip (i.e., toward the tread). Judging the general ozone deterioration to the Shell rating scale on the whitewall side of the tire also must be done in the upper sidewall and buttress area beyond the location of the cover strip.

- **The process of ozone deterioration takes time to form**—Therefore, sidewall and tread cuts, gouges, scrapes, impacts, and so forth with ozone cracking or crazing across these damage areas indicate that time has passed, and they are not newly made. On the other hand, when rubber with ozone deterioration cracking already present suffers a cut, snag, gouge, or so forth, the cracking lines will be somewhat distorted through the rubber.

[4-27] Usually, the outboard side can be determined by the degree of cleanliness of the sidewall. (No one washes the inboard side of the tire.) However, the degree of ozone deterioration also is fairly definitive.

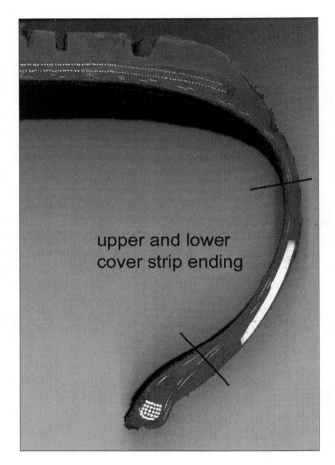

Fig. 4.49 A whitewall tire section, indicating the ending of the cover strip.

Within all the categories discussed in this section, awareness and notation should be made of any changes in the levels of ozone deterioration, either spot changes or sidewall-to-sidewall variations. Dramatic changes in the amount of cracking can occur with improper storage, man-made (local) ozone creators, and high and consistent exposure to the sun.

Ozone deterioration in the tread and specifically the tread shoulder can lead to crack depths that approach or reach the belts (Fig. 4.50). Cracking through the tread or buttress that approaches the belt compound allows ozone to be in contact with the belt compound, beginning the process of mechanical and chemical degradation of that compound. Degradation of the belt compound from ozone can cause belt separation.

Identification of Causes and Contributors

Fig. 4.50 Ozone deterioration to the belt wires (7x magnification).

4.6 Mounting and/or Demounting Damage

Mounting or demounting damage (Fig. 4.51) from a mounting machine will appear as a cut or tear [Ref. 4.5, pp. 142–145] that generally will have a parabola shape on the bead face. The tear itself may or may not appear to be parabolic in nature, but if you look closely (sometimes magnification is required), an indentation into the bead face compound of the mount or demount tool can be seen riding up from the bead toe across the bead face[4-28] and then back down to the bead toe again. Depending on the type of machine and condition of the tool, the tear can be in any location in the bead toe or bead face area, and the indentation

---

4-28 See Fig. 4.55 later in this chapter for the locations of the bead toe and bead face in a tire.

Fig. 4.51 Bead tear mounting or demounting damage.

into the rubber on the bead face may form only a partial parabola. Pure accident damage to the bead will tend to have a complete randomness to its appearance (Fig. 4.52), there may be multiple locations of the same type of damage, or the damage will include the bead toe, face, or heel, but also the part of the bead that rests against the wheel flange and/or the lower sidewall.

Mounting or demounting damage using hand tools will leave small impressions and tears in the bead toe, bead face, or bead heel. They will not have the parabola appearance of a machine mount, and typically there will be more than one bunched together.

Mounting and/or demounting damage that results in a belt separation indicates that the bead toe or bead face has been torn and that the tear has exposed the ply cords to the internal air pressure, causing ICP.[4-29] (See Section 4.1.2 for more

---

[4-29] This does not include large tears that result in detached beads and/or excessive rocking on the rim, which can lead to more immediate ride and vibration problems and subsequent tire disablement, with or without belt separation.

Identification of Causes and Contributors

Fig. 4.52 Bead tear accident damage.

details on ICP.) Most accident damage to the bead area is caused by the wheel during an accident sequence. However, there are times when a wheel flange (damaged or undamaged) can ride up the bead toe, mimicking mounting or demounting damage.

To identify wheel flange damage versus mounting or demounting damage, magnification of the area is required, as well as an examination of the wheel. With 4x to 7x magnification, a wheel flange cut or tear will tend to resemble a knife-like cut or a pull-apart tear. Most, but not all, mounting or demounting damage in the initial formation of the tear will tend to be a compression tear. If the wheel flange is not damaged and there are no sharp areas, it is unlikely that the wheel was the cause of the damage. In rare instances, wheel flange accident damage appears, on the surface, to resemble mounting or demounting damage. However, magnification of the area again will reveal the type of tear.

Each bead tear should be pulled open and the inside of the tear inspected for exposed ply cords, deep tears to the ply compound, dirt, and grime. Superficial tearing affecting only the fabric or rubber chafer can be noted in the inspection, but it is unlikely that those tears will cause ICP.

The inside of the bead tear can reveal time-dependent events, as follows:

- **The interior of the tear appears clean and fresh**—This could be a recent cut, a tear in the post-accident demounting of the tire from the wheel, a mounting or demounting in the recent past, or damage from the accident sequence itself.

- **The interior of the tear has dirt, grime, or evidence of rubbing together within the tear**—This indicates that the tear is not "fresh" and could be from a pre-accident mounting or demounting that is not from the recent past. During the accident sequence, it is possible for dirt and grime to enter the bead tear. However, for this to occur, the tear needs to be opened, and there should be fresh abrasions or marks on the bead surface of the tear, indicating an opening event.

- **The tear has both fresh tears and dirt and grime**—This leads to the conclusion that some damage occurred during the demounting post-accident or during the accident sequence, but there also was damage to the bead pre-accident and not in the recent past.

In Fig. 4.53 the tire bead was torn completely off during demounting prior to inspection. (The torn piece has been replaced to its original location in Fig. 4.53.) However, there was a previous tear at this area, as indicated within the white squares of Fig. 4.54 and noted by the dirt accumulation.

Fig. 4.53 A bead tear from demounting, post-accident.

Discovery evidence can be helpful and telling as to when the tear occurred. Regardless of whether it is viewing of the accident scene, storage lot photos, or deposition testimony, the final determination of cause of the bead tear should be made using all available evidence.

Some final comments should be made on bead tear. The movement of the bead area during normal use is a rocking motion around the bead core (Fig. 4.55). Due to this motion, a bead tear of sufficient depth to allow movement will show some

Identification of Causes and Contributors

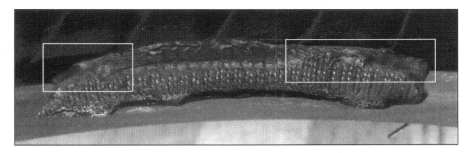

Fig. 4.54 A bead tear from Fig. 4.53, indicating pre- and post-accident tears.

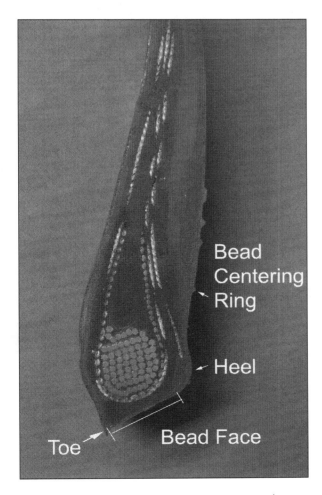

Fig. 4.55 Bead area of a two-ply light truck tire.

wear in the interior, depending on the exact location of the tear, the number of miles driven, and over-deflection. An area directly below the center of the bead core over to the bead heel will not show much movement or working because this is a fairly stationary area of the bead. The closer the tear is to encompassing the mid-bead face to the bead toe, the more movement there will be.

Tears that are from the bead heel upward into the bead flange area and bead centering ring are generally not involved with ICP because tires normally seal off the air from the bead toe to the bead heel. However, tears from the bead heel to the atmospheric air can lead to leakers.

Finally, bead tears can be investigated using time-dependent events, including other rubber indentations and movement that occurs naturally over time, such as compression grooves or circumferential "lines" in the middle of the bead-face (see Section 4.2 on over-deflection), or damage and dirt locations from the accident sequence. When investigating time dependency, keep in mind which item overlays the other. Whatever was laid down first in the history of the tire will be overlaid by what came second.

4.7 Physiological Aging[4-30]

Aging is a phenomenon of all things on earth, whether it be tires, people, pets, or nuclear fuel. Determining the chronological age of a tire is as easy as reading the DOT code. Determining the physiological[4-31] age of a tire is more difficult, but several methods can be used without the involvement of destructive testing.

4.7.1 Durometer[4-32]

On a molecular level, as tires age, the number of cross links[4-33] within the rubber will increase, and sometimes, to a lesser degree, chain scission occurs. The cross-linking increase raises the hardness values.

[4-30] As of this writing, there is no government aging test requirement for tires. The National Highway Traffic Safety Administration is developing an aging test for tires.

[4-31] Note that this is not the chemical deterioration of the tire, which would be addressed by a chemist or chemical engineer, and is not covered in this book.

[4-32] Modulus (hardness) of the compound is measured with a non-destructive tool called a durometer. This device, for rubber, measures the hardness on the Shore A scale using a range from 0 to 100, with 100 being the hardest.

[4-33] Cross links are molecular covalent bonds linking polymer chains.

Identification of Causes and Contributors

Thus, if you were to take the same tire and measure its hardness a few days after it was cured and again each year for 10 years, there would be an incremental increase in the durometer value over time. Each tire manufacturer will have different new-tire hardness values for its components. For all-season tires, reasonable durometer values for new tire tread range from 55 to 70 and for sidewalls from 55 to 65. For high-performance tires and ultra-high-performance tires, the values can be both low (less than 50) and high (greater than 70). Therefore, judging the physiological age of a tire based purely on a durometer alone requires experience. However, tires normally show multiple determinations of oxidative and ozone deterioration and are not solely durometer-based.

During the inspection process, treads and sidewalls should be checked for durometer values.[4-34] Measurements of the tread should be taken in at least four places around the circumference of both tread shoulders and their values given as the minimum and maximum recorded. Sidewall measurements must be measured in the upper sidewall area on both sidewalls. The flexible sidewall will have to be supported physically from the inside with one hand to get a proper value. If one sidewall is a white sidewall, ensure that the measurement is not taken in the whitewall nor in the cover strip.

With a classic hand-held durometer, it is impossible to obtain values of thin rubber components such as belt compounds, innerliners, ply compounds, and so forth from a cured tire, even after tread/belt detachment.

4.7.2 Appearance and Feel

In addition to the preceding hardness numbers, the appearance and the feel of the sidewall, liner, and tread compound versus the chronological age of the tire are as follows:

- **Appearance**—The appearance of the sidewall and tread compound should be such that the rubber is not abnormally cross hatched (Fig. 4.56) with ozone-type cracking.

- **Feel**—The feel of the sidewall compound, when bent with the fingers, should have elasticity to it versus a brittleness or hardness. This also is true of the tread lugs when stretched apart with a tread block spreader. [4-35]

[4-34] To obtain a proper measurement, do not measure close to the siping and tread block edges, which softens the value (i.e., makes it less).

[4-35] This obviously takes experience because there is no tactile elasticity standard.

Fig. 4.56 Abnormally cross-hatched sidewall surface.

The liner elasticity is judged tactilely and visually by pressing into it with the thumb and observing it when rolling the tire on a hard surface on one shoulder and the other shoulder with some force such that the liner is bent somewhat inward.

4.7.3 Spot Ozone Damage

Tires that are stored improperly or exposed to consistent ozone sources can have a portion of the tread with a substantially higher durometer than the remaining portions of the tread. It also is possible that one sidewall might have a durometer value that is much higher than the value for the opposite sidewall. Furthermore,

Identification of Causes and Contributors

the sidewall also can have a portion of its area with a much different value than the other areas of the same sidewall (Fig. 4.57). Spot ozone damage usually occurs from poor storage, on or off a vehicle (i.e., the same side of the vehicle exposed to the heat/sun, spare tire storage problems). Although the varying durometer values[4-36] within a tire can be indicative of improper storage, the feel and visual appearance of the sidewall also will show a difference if the durometer values are substantially different.

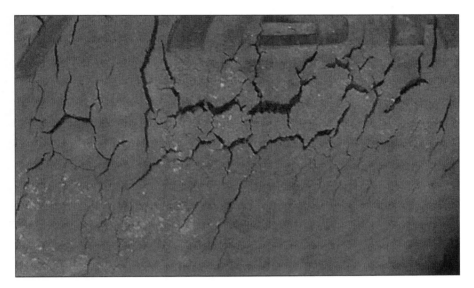

Fig. 4.57 Spot ozone damage of the sidewall.

4.7.4 Belt Tearing

As mentioned in Section 4.1.2 on ICP, near-circumferential belt rubber tears, as opposed to radial or angular tears in the belt compound, can be a sign of oxidative damage. Intra-carcass pressurization should always be investigated if this kind of tearing is seen in the belt compound. A good visual appearance of the belt should have neither a cracked nor a crazed appearance. However, this needs to be judged also by keeping in mind the time from the tread and belt detachment to the inspection, as well as any chain-of-custody issues.

[4-36] Care must be taken when using a hand-held durometer. Variability in the use of the tool is dependent to some degree on the person taking the measurements. Most tire forensic experts will come within 2 or 3 points of each other on the range.

4.8 Snags, Gouges, Cuts, Tears, and Abrasions

Some tires will tell a story simply by their appearance. In Fig. 4.58, all the sidewall markings on one side of the tire have been abraded off, except for the lowest tire stamping near the bead centering ring. Obviously, this tire spent a lot of time scuffing against objects, most likely curbs. How that affects the causation determination of the failure should be investigated, but it is part of the damage history of the tire.

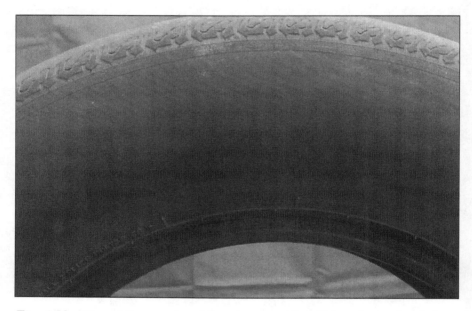

Fig. 4.58 All markings on the mid- and upper sidewall have been abraded.

If a tire sustains damage during the accident sequence in an uninflated state, the determination of damage areas prior to and after the beginning of the accident sequence hangs on the oddity or the regularity of these items.

If damage found in the tread or sidewall has regularity[4-37] (e.g., the sidewall abrasion in Fig. 4.19), this generally indicates a pre-accident event. When a tire

---

[4-37] Here, regularity means spacing or type or in the consistency of the radius or angularity from the center of the wheel.

Identification of Causes and Contributors 103

is uninflated during the accident event, it is difficult to be damaged in a regular sequence. If the tire is inflated, regularity in damage can occur. Because it is not unusual for tires to suffer a tread and belt detachment and to remain inflated, the timing of the deflation (if any) should be investigated versus the damage found, including the amount of "road rash" observed.

The "oddity" can occur when it appears, for example, that the accident damage is directional but something appears in the opposite angle or direction, or all the damage is on the outboard side of the tire, but a sole mark appears on the inboard side (Fig. 4.59). Because the oddity can include thousands of examples, these two examples are not meant to be the entire universe of oddity sets. One should have a basis for the oddity being causative, which derives from the inspection of the tire and from documenting what is seen and felt. (See Fig. 4.60 for a sidewall impact mark.)

Fig. 4.59 A sidewall mark at the mid-sidewall point.

Regardless, if the damage is regular or an oddity, the angularity and direction (i.e., bead to tread, or tread to bead) should be noted. The direction of the damage (Fig. 4.60) is determined by the "lines" that form in the rubber from the striking object.

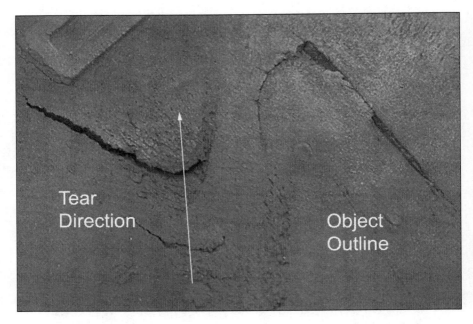

Fig. 4.60 A complex directional tear with an imprint of an impacted object.

When rubber is pushed by an object (i.e., abraded), the object will raise lines in the rubber. The movement of this object from its contact with the rubber to its ending point is perpendicular to the lines. Unless the tire was used off-road, the direction of damage from objects in, on, or under the road surface generally is from the tread to the bead, whereas accident damage from the wheel generally is from the bead to the tread (typically wheel damage).

Tears, cuts, snags, abrasions, and gouges that reach or nearly reach the belts, plies, or bead reinforcements all potentially can cause tire failures. The quantity, type, depth, location, and direction of tears, cuts, snags, abrasions, and gauges also can say something of the "normal" working environment that the tire experienced.

4.9 Cutting and Chipping

Cutting and chipping (C&C) refers to the tearing and gouging (Fig. 4.61) of the tread area of a tire. A moderate-to-severe cutting and chipping of the tread surface amounts to hundreds or thousands of small impacts. This type of damage could be referred to as "death by a thousand cuts." It is rare, but quite possible, for

Identification of Causes and Contributors 105

Fig. 4.61 Cutting and chipping of the tread.

cutting and chipping to be the primary cause of a tire failure. However, cutting and chipping can lead to tire failure in the following manner:

- A cutting-and-chipping belt separation and/or environment is a prime penetration and puncture environment. These environments are non-improved road surfaces such as gravel or dirt roads. As the cutting and chipping gets closer to the belts, the chances of penetrations, punctures, and belt damage increase. (See Section 4.1 on punctures and Section 4.3 on penetrations for more explanation.)

- As the cutting and chipping approaches the belt package, additional stress is added to the belt edges, which can lead to crack initiation in the rubber. This can lead to direct tire failure or tread belt detachments.

- With decreasing tread depth from cutting and chipping, ozone will have the ability over time to work its way into the belt compound, causing belt separation. (See Section 4.5 for more details about ozone deterioration.)

There is no bright line standard for a general cutting-and-chipping scale. An easy system, but based on experience, is as follows: slight (i.e., some cuts, a few small chunk-outs); moderate (cutting on most lugs, surfaces, and chunk-outs regularly seen, some of which can be deep); and severe (substantial chunk-outs and cutting of the tread with regular deep cuts). Moderate cutting and chipping is by no means acceptable unless we are talking about commercial tires that are used regularly in a cutting and chipping environment.

Sometimes the environment to which a tire is exposed changes over time. Although the sides of the lugs and in particular the bottom of the grooves and slots do not come in contact with an improved road surface, the wearing surface of the tread lugs or ribs does come in direct contact with improved road surfaces. Therefore, at times, the tread surface might be somewhat smooth with only a little cutting and chipping, but the sides of the lugs or ribs and the bottoms are torn, cut, and chipped to a much greater extent. This would be an example of a tire first exposed to a cutting-and-chipping environment and then to an improved road surface environment.

4.10 Poor Tire Storage and Improper Tire Maintenance

4.10.1 Poor Tire Storage

Poor tire storage has been covered in some detail in Sections 4.2, 4.5, and 4.7.1 under the discussions of over-deflection, ozone deterioration, and durometer values, respectively.

Although electric motors and welders create ozone and may cause "spot" damage to a nearby tire, storage in an area that is open to sunlight for extended periods of time can cause ozone deterioration on one side versus the opposite side of the sidewall. Several examples of poor storage conditions are as follows:

- **Crushed or stored without air**—Storage off the vehicle or on the vehicle (such that the tire is crushed radially or circumferentially unmounted or mounted, without air pressure, for extended periods of time) will show evidence of creasing or cracking, even after re-inflating the tire. For instance, a tire that is stored on a vehicle without support or air pressure (e.g., a flat tire) may come apart at a later date, depending on the amount of time that the tire was flat and the surrounding environmental conditions. The creasing or cracking is the ozone attack of the compound at a high stress point (i.e., the folded or crushed parts of the tire).

- **Stored in contact with petroleum-based products**—Storage where petroleum-based products come into contact with tire rubber will cause a chemical

Identification of Causes and Contributors

breakdown in the polymers. The breakdown will have the characteristics of significantly lower durometer values than normal, with softness and sponginess to the touch. The rubber will have a porous texture that will be visible with the naked eye or with slight magnification. Finally, unless much time has passed since the contact with the petroleum-based product, there may be a scent of "oil." The extent of the time of contact with the chemical can cause tire conditions to range from poor wear to a tire failure.

- **Storage in such a position that the tire experiences spot heating**—Localized high temperatures will prematurely degrade the tread and belt stock (increasing its hardness) from the surface inward. This part of the tread will have significantly higher durometer values than other parts of the tread.

These examples are by no means the range of poor off- or on-vehicle storage. Further reference material is available at all major tire manufacturers' web sites.[4-38]

4.10.2 Improper Tire Maintenance

Improper tire maintenance includes, but is not restricted to, air pressure and loading not maintained per the vehicle placard or the owner's manual, not inspecting the exterior of the tires regularly, and not following the vehicle manufacturer's instructions on tire rotation, vehicle alignment, and balance. Other examples would be improper or poorly made repairs and improper size, type, or placement of tires on the vehicle, as well as improper usage of the tire.

Generally, improper tire maintenance is determined by the visual appearance of the tire. This is discussed in detail in other sections of this chapter. In addition, Tire Industry Association (TIA) manuals [Refs. 4.5 and 4.15], the Rubber Manufacturers Association (RMA) "Tire Care and Safety Guide" [Ref. 1.5 in Chapter 1], the National Highway Traffic Safety Administration (NHTSA) "Tire Safety: Everything Rides on It" [Ref. 4.16], and pamphlets at most local tire dealers and tire manufacturers inside new vehicles and at the web sites all contain information covering the proper care and maintenance of tires.

Identification of the various types of irregular tread wear and the possible causes for that malwear produced from poor tire maintenance are important in eliminating or identifying possible causes for tire failure.

[4-38] For example, Yokohama technical bulletins can be found at http://www.yokohamatire.com/pdf/tsb-112102.pdf.

Irregular wear types from poor tire maintenance include the following:

- **Center wear**—The center of the tire wears faster than the shoulders.
- **Shoulder wear**—The shoulders of the tire wear faster than the center.
- **Heel and toe wear**—The tread elements wear in a saw-toothed pattern.
- **Rib depression wear**—The intermediate ribs wear faster than the outside ribs. This is seen mostly in commercial application light truck tires or medium/heavy truck tires.

4.11 High Speed and High Ambient and/or Pavement Temperatures

4.11.1 High Speed

Because of centrifugal force, running tires at higher speeds[4-39] than their capabilities increases the stress and strain within the belt edge region, which then builds high belt-edge temperatures and distortions in the geometric shape as the tire expands in diameter, including the onset of the standing wave [Ref. 1.1, p. 409].

The destruction of the tire from the standing wave effect will indicate a mechanical-type failure, with tension-caused tears. Freely spinning tires that are not in contact with the road surface (i.e., when one tire of a drive axle of a vehicle is stuck in a snow bank and the other tire is allowed to spin free) also can fail [Ref. 1.8, p. 8] by tension tearing. Speed-caused tire failures can be explosive; therefore, there will be tensile breaks in the wires, tread and belts torn off each other, and usually casing rupture in tension. Pure speed-related failures are rare and are fairly easy to diagnose from the tensile-type breaks. Likewise, they usually are accompanied without the classic fracture surface of a belt-to-belt separation.

It is quite possible to have bluing of the components of the tire due to heat from speeds that are less than the speed code of the tire due to under-inflation or overload. The area of the belt edge also is in the tread shoulder, which in passenger car and light truck tires usually is the area with the greatest thickness and volume of material. As a tire heats up from increased speed in an over-deflected

---

[4-39] In a pure sense, the definition of what constitutes high speed is not only speed itself but time and is dependent on the tire service description or other speed-limiting factors if they exist. For tire forensic purposes in this book, unless stated otherwise, high speed or "at speed" will be used to indicate highway-type speeds, which are 88 to 137 kph (55 to 85 mph).

Identification of Causes and Contributors

condition, the belt edge area, due to its greater thickness, will not be able to dissipate the heat fast enough to maintain a constant temperature. This continuing rise in temperature then will begin to degrade the belt compound properties, possibly leading to belt separation.

4.11.2 High Ambient and/or Pavement Temperatures

High ambient and, coinciding with that, high pavement temperatures that affect tires are found primarily in the southwestern states. Although these items may not be the primary causes of a belt separation, they certainly affect the running temperatures of a tire and thus the belt temperatures of a tire. High belt edge temperatures are a known factor in belt separation, especially when mixed with other causes of belt separation (e.g., over-deflection, poor storage, impact, punctures). It is not likely that an examination of a tire results in a statement that high ambient or pavement temperatures were involved in the tire failure without further knowledge, most likely from discovery evidence. However, it is quite possible to state that heat was involved by bluing conditions and physiological evidence in the various components that indicate a high heat buildup and/or a history of the environmental condition of the tire.

4.12 Vehicle-Related Conditions

Items in this grouping are numerous because the tires are connected directly to the vehicle and the road. These items include the following:[4-40]

- Alignment conditions (e.g., toe, caster, camber)

- Poor vehicle suspension component maintenance (e.g., struts, shocks, springs)

- Wheel, hub, or bolt circle damage

- Hard cornering and braking

- Misapplication of the rim to the tire or the tire to the vehicle (including duals)

[4-40] Tire Industry Association (TIA) manuals (Refs. 4.5 and 4.15) include photographs, descriptions, and examples of the various vehicle-related conditions and the irregular tread wear patterns that can develop.

Belt separation may occur with each of these items when the condition causing the tire to wear irregularly is not noted and is allowed to continue until the tire develops a belt separation (Fig. 4.62). For instance, camber, and especially negative camber (i.e., wearing the inside shoulder), can go unnoticed. Then, as the one-sided wear progresses through the tread, a belt separation can form, or a sudden air loss may occur with continued use of the tire, and the tire is worn through the casing.

Fig. 4.62 Fast wear on the outside shoulder, with underlying belt edge separation.

Tire conditions caused by the vehicle are seen mainly in the type and amount of abnormal tread wear. Judging abnormal wear (sometimes called malwear or irregular wear) caused by vehicle-related conditions is important because abnormal wear causes abnormal forces to be added to the existing stress and strain of the belt edges. These forces can lead directly to belt separations.

As the tire tread wears down toward the belt package, the forces on the belts increase, compared to when the tread is at full depth. When a mechanical condition in the vehicle causes a fast wear on one or both shoulders, the belt edges will have higher stress levels than the non-fast wear shoulder and can develop a belt separation from the alignment condition alone. This is not the same as an accelerated wear area from a belt separation.

The mechanical condition of brake drag or hard braking down steep slopes can create heat, which can degrade the steel ply and belt properties. The heat generated from this condition can be intense and can cause tire fires (almost exclusively in commercial light or medium/heavy truck tires).

CHAPTER 5

Identification of Non-Belt Separations

5.1 Tread Separation

A tread separation can occur between the top side of the top belt and the tread [Chapter 4, Ref. 4.5, p. 42–45], sub-tread, or tread cushion after the tire is in use.[5-1] A tread separation is not a tear of the tread lugs or the ribs off the belt from some outside force. If the tread is torn off the belt, then that is considered a tread tear rather than a tread separation. This difference can be determined by investigating whether the separation is interfacial (i.e., smooth), with multi-planar tearing, or with degraded rubber tearing.

A tread separation off the top belt that does not involve a belt separation, puncture, poor repair, impact, or similar factors can be due to contamination of the top belt surface or the under-tread surface. If the tread is separated from the top belt and the separation encompasses a belt-to-belt separation, the area should be examined carefully and a determination made that the separation did or did not start at the belt edge and go over the top of the belts. If the tread separation encompasses the belt edge, it might likely be an extension of the belt separation and not a true tread separation. If contamination is suspected, chemical testing[5-2] might be required

---

[5-1] Prior to the placement of the tires in service, this type of "separation" would be considered a curing condition.

[5-2] Reference 4.17 in Chapter 4 contains a general discussion by Dr. J. Rancourt of Polymer Solutions Inc., Blacksburg, Virginia, on the types of chemical testing used in the analysis of tire failure surfaces.

to identify the cause. However, remember that once a separation is open to the environment, contamination from the environment can affect the results.

5.2 Bead Area Separation

A bead area separation[5-3] can be the lower sidewall separating off the plies, a separation between the ply turn-ups or turn-downs, a separation of the bead bundle from the surrounding plies, or a separation between the fabric or steel chipper/chafer around the bead.

Bead area separation in passenger tires has been fairly uncommon since the middle to late 1990s. In light truck-type tires, especially load range D and E sizes, bead area separation is unusual but not rare and remains small in number compared to the number of belt separations. Regardless of the rarity or non-rarity, it is necessary to identify this type of separation versus indentations that are not a cause for concern. For example, Fig. 5.1 shows an indentation in the sidewall due to component lay-ups and typically is not a concern for tire failure.

On a mounted and inflated tire, a bead area separation, unless it is caught early in the form of a bulge[5-4] or a circumferential crack (usually in a near-constant radius) in the lower sidewall, typically will result in sudden air loss and a subsequent run-flat condition.

Flow cracks[5-5] also tend to be circumferential and usually on a near-constant radius, and they can be found in any location in a tire. Figure 5.2 shows a flow crack on the bead face. Depending on its depth, location, and reason for being in the tire, a flow crack can be harmless or detrimental to the tire. In most cases, flow cracks are harmless due to the nature of their existence. However, if flow cracks occur in an area of high strain, they can form cracks that go deep into the tire.

Because the appearance of flow cracks can mimic a cut along a defined radius, one must inspect this area closely to make the determination that it is a flow crack.

[5-3] A bead separation occurs in the bead area, which includes the lower sidewall, ply and lower reinforcement endings, bead flange, and bead face areas.

[5-4] Some bulges relate to the placement of underlying components and the way they are laid up in the building state. These bulges usually are not a cause for concern.

[5-5] Flow cracks occur during vulcanization when rubber components move and try to fill a void, and either there is too much or not enough rubber to do so and/or a surface condition prevents them from "knitting" together, giving the cured appearance of rubber flow.

Identification of Non-Belt Separations

Fig. 5.1 A 360-degree indentation in the lower to mid-sidewall.

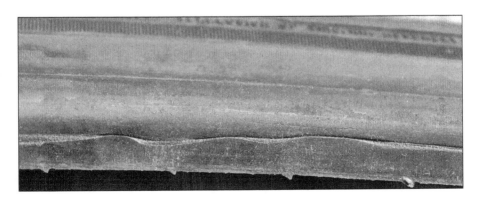

Fig. 5.2 A flow crack in the bead face.

That determination will involve some knowledge of tire design, noting that the appearance of the "crack" near the surface is more rounded than a cut.

5.2.1 Lower Sidewall Compound Separation Off the Plies

This type of separation occurs on the outside of the plies and might be visible in the form of a bulge (mounted and inflated) or a tactile soft feel when squeezed (unmounted). Over-deflection, intra-carcass pressurization (ICP), severe torque, and rim pinches can cause this separation, as can foreign material and misplaced components. If the area has cracked open[5-6] and the separation can be viewed, the type of separation should be noted. Medium magnification can reveal if the separation is interfacial or a ragged tear type of separation. Ragged tears in this area will be very small; however, an interfacial separation between the sidewall compound and the ply compound can be quite smooth and sometimes shiny (e.g., similar to trapped air).

5.2.2 Separation Between the Ply Turn-Up(s) or Turn-Down(s)

This type of separation will appear as a bubble or as softness when squeezed with the fingers on an unmounted tire or as a circumferential crack in the lower to mid-sidewall of the tire. From the innerliner side of the tire, this separation can appear as a bubble or a circumferential crack upward from the bead toe in the lower sidewall.

Separations between the upward or downward plies can be caused by over-deflection, excessive torque (although an angular crack in the mid- to upper sidewall will appear first), excessive high speed, rim pinches, foreign material, or improper component locations.[5-7]

[5-6] Remembering that this is nondestructive failure analysis, no cutting of this area should be attempted without approvals from counsel.

[5-7] Improper component endings generally are from the tire building process. An extensive knowledge of tire design is required to make this judgment.

5.2.3 Separation of the Bead Wires or Bundle from the Surrounding Plies

This type of separation is rare and would be unlikely to cause a tire failure, except in cases of high over-deflection, when the separation can expand upward into the ply endings. Loose bead wire can "walk"[5-8] its way out of a tire after breaking loose from the plies.

The cause of this type of separation generally is from poor adhesion between the wire bundle and the surrounding fabric. A loose bead bundle without loose wires is rare and usually appears in combination with another lower sidewall separation. In general, however, the bead core is highly inert to movement and separation, even in tires that accumulate substantial damage over time.

5.2.4 Separation Between the Steel or Fabric Chipper[5-9] or Chafer

Separations in this area are in the lower sidewall and can manifest themselves visually, externally as bulges, as circumferential cracking, or as softness when squeezed with the fingers. Internally (i.e., on the liner side), they can appear as bubbles, bulges, or cracks. Because they are reinforcement agents to the lower sidewall and not part of the main structure, the separations from these components will not necessarily cause a sudden air loss, unless the separations expand over time into the ply endings or actually "work" their way out of the external lower sidewall.

Over-deflection, impacts, excessive torque, component placement, and foreign material can be causes of these types of separations. To define how these separations occurred, inspect for evidence of lower sidewall damage, the quantity of over-deflection that is present, internal separations (if possible), ragged tears, bluing of the ply stock, melted ply material, or smooth interfacial rubber separations.

5-8   For the wire to "walk," the bead must be loose 360 degrees.

5-9   Although the terms "chipper" and "chafer" are used interchangeably, a chipper generally is an outside-only reinforcement, whereas a chafer will wrap the bead on both sides. Note that in truck and bus tires, a chipper can wrap the bead.

In rare cases, a belt separation can lead to a bead separation in the same area as the belt separation, or vice versa. This is due to the additional strain that occurs on the non-separated belt or bead when the belt or bead reinforcement areas begin to come apart. If the belt separation and the bead separation occur at almost the same radial location, it will be necessary through time-dependent events to determine which came first and, of course, if possible, why.

5.2.5 Bead Breaks

Broken beads are uncommon but can fail in the following instances:

- A mismatch of the tire and the wheel, such as mounting a 16-in. tire onto a 16.5-in. wheel

- Improper mounting methods or obstructions causing a "hang-up" of the bead in the wheel well

- Shipping damage to the bead area

- Tires mounted with kinked beads, indicating prior bead damage

- Insufficient bead strength (i.e., not enough wires), incorrect wire yield strength, incorrect wire type (ultimate tensile), or incorrect wire diameter

- Incorrect (low) bead winding diameter and/or an incorrect (high) bead seat diameter

During the investigation of a bead core break, the wires should be analyzed under magnification to determine the type of break (Fig. 5.3). Certain breaks could indicate damage from outside forces having nothing to do with any of the previously mentioned bead break items. Creel-type beads (also called weftless or tape beads) will have an overlapping splice on the bottom and top of the bead bundle. Beads failing in pure tension will break at this splice. Identification of the splice can be accomplished with 15x magnification and noting that the bottom or top strands of the wires are cut while the remainder of the bundle shows tensile breaks. (The cut wires are from the manufacturing process and are normal.) Also, the splice overlap can be identified by counting the individual wires on one side of the break versus the other. One side will have more wires if the break is at the edge of a lap splice.

Identification of Non-Belt Separations

Fig. 5.3 Tensile breaks of the bead wires.

5.3 Sidewall Separation

A sidewall separation is a separation between the plies in the mid- to upper sidewall area, a separation between the sidewall compound and the ply compound, or a separation between two sidewall compounds (i.e., white sidewall components).

5.3.1 Separation Between the Plies

A separation between the plies in a multi-ply radial tire is a rare event. It is most closely associated with bias and bias belted tires. Even in radial tires with ply angles slightly off radial (e.g., 87 degrees), a separation between the plies is rare. These separations will appear as bulges or softness in the sidewall. Unless they have cracked open to allow a good inspection, cutting open the sidewall[5-10] is the only method to inspect the interior sections. (X-ray and shearography will have some usefulness in identifying the area but will not identify the causes.)

[5-10] This is considered destructive testing and should not be attempted without permission from the litigants.

True separations between the plies and not accident damage, road hazard impact, or over-deflection-induced ply compound tears are so rare that manufacturing conditions should be investigated as one possibility.

5.3.2 Separation Between the Sidewall Compound Components

The white sidewall portion of a tire is made up of several layered rubber components. In contrast, black sidewalls may be made from one to three compounds. Small contained separations within these components do not cause tire failures. Furthermore, they appear as blisters when near the surface or a rounded raised area, and they most likely will be related to trapped air.[5-11] Trapped air can easily be identified if the area can be cut open. A very smooth shiny surface against the casing or between the rubber components is evidence of trapped air. Small amounts of trapped air generally do not cause separations.

During a tire failure or accident sequence, the whitewall can become detached from the sidewall. It is highly unlikely that the whitewall came off the tire first; most likely, it came off during the accident sequence. This is due to the nature of the middle third of the sidewall region. While there is a great deal of movement in this area during tire rotation under load, as long as the rubber components are designed appropriately and layered properly, it is a part of the tire without much in the way of stress risers. Without stress risers, the chance of separation is very small.

5.3.3 Separation Between the Ply and the Sidewall Compound

This type of separation comes in only two forms. The first is an interfacial separation, which is a smooth type of separation. The second is a multi-planar separation within the ply or sidewall compound. Due to the thinness of the ply coat stock and the rippling effect of the ply within the rubber matrix in the mid- to upper sidewall, magnification might be needed to differentiate between the two. The causes of this type of separation can have multiple sources, including punctures, breaches of the liner, impacts, or foreign material.

[5-11] Trapped air will be due to the building process.

CHAPTER 6

Identification of Various Tire Conditions

6.1 Run-Flat Damage[6-1]

Run-flat damage is caused to the tire after the tire has lost its internal air pressure and the tire has been "run flat" by the driver. This is different from damage caused by a sudden air loss that occurs with or without a tread/belt detachment. In a run-flat tire, the damage is caused when an intact tire has zero or near-zero air pressure, and the user has continued to run on the tire either unknowingly or on purpose (e.g., to reach the next exit or a safe stopping area). Continued use of the tire in a run-flat condition quickly creates high heat within the tire and high shear strains in the upper sidewall. These forces eventually will shear the sidewall from the tread (Fig. 6.1).

Evidence of run-flat conditions can be as follows:

- Multiple radial casing breaks in the sidewalls (Fig. 6.2), with possibly little to no belt material remaining. This condition is more common on axles with dual tires or tires with steel body plies. In the dual condition, the tire with air carries the load, while the tire without air continues to shred and shed pieces of rubber and reinforcement.

- Severe casing tearing, cuts, and gouges (more likely in dual-tire arrangements).

[6-1] See Appendix D for a pictorial representation of an actual run-flat occurrence.

Fig. 6.1 Run-flat damage with only two sidewalls remaining.

Fig. 6.2 Lower sidewall run-flat damage—multiple radial casing breaks.

Identification of Various Tire Conditions

- Sidewall(s) that have been severed from the tread and belt package (Fig. 6.1) up to 360 degrees.

- Sidewalls that have been severed above the bead up to 360 degrees.

- Mild to severe abrasions 360 degrees in the upper to middle sidewall on the exterior of the tire.

- Mild to severe abrasions in a circumferential direction on the interior (i.e., liner side) of the tire from the lower to the upper sidewall (Fig. 6.3).

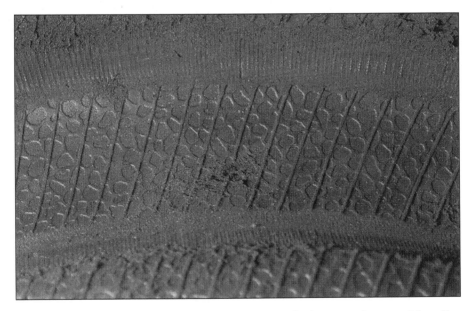

Fig. 6.3 A run-flat condition—liner abrasions in the lower and upper sidewalls.

For a tire to be termed run-flat, any one or a combination of these conditions must occur with an intact tread and belt package or no belt package supplied for inspection, in the case of severed sidewalls. Run-flat tires can have belt separations within the tread belt package; however, there will be no tread belt detachment in a run-flat tire. If a failure is caused by a belt separation, then the failure is a belt separation rather than a run-flat condition.

On recognizing this type of failure, one needs to find the cause of the air loss. Due to the amount of damage sustained during a run-flat, the cause may not be found, even after painstaking reconstruction of the sidewalls, the beads, and the intact tread. However, this type of failure definitely is from an air loss and typically is not a defect in the tire. The following are several potential causes:

- **Puncture**—This refers to a non-repaired or poorly repaired puncture.

- **Road hazard impacts**—These are impacts with objects on, in, or under the road.

- **Rim pinch**—This is a derivative of a road hazard impact and is a circumferential or half-moon-shaped mark above the bead centering ring in the sidewall. Typically, a rim pinch may be low in the sidewall (Figs. 6.4 and 6.5) or high (less common) in the sidewall.

Fig. 6.4 A rim pinch, viewed from the exterior. Here, an impact led to a run-flat.

- **Bad valve**—The wheel valve should be checked by a backpressure test to test the valve integrity at the rated pressure of the wheel or secondarily at the maximum pressure listed on the sidewall of the tire.

- **Wheel**—The wheel may be cracked or damaged.

- **Vehicle alignment conditions**—These conditions are affected especially by camber, as can be seen in Fig. 6.6.

Identification of Various Tire Conditions

Fig. 6.5 A rim pinch of Fig. 6.4, viewed from the interior.

Fig. 6.6 A run-flat through one sidewall.

- **Torque cracks**—These cracks indicate a breach of the liner in the sidewall from a combination of over-deflection and torque (Fig. 6.7).

By definition, in a run-flat tire, the tread and belt package were intact prior to the loss of air; therefore, the causes of failure that involve portions of the belt and tread

Fig. 6.7 Innerliner cracks off angle from the ply angle.

integrity should be attributed to other failure causes. At times, the only evidence received are the two lower sidewalls, without the tread or the wheel to inspect. By looking at the lower sidewalls, enough evidence can remain to declare the tire a run-flat, even if the root cause of air loss cannot be determined. Many tires that have two lower sidewalls remaining will have 360-degree circumferential abrasions in the upper sidewall that came about prior to the sidewalls severing from the intact tread.

Signs of short-term high over-deflection may appear as the tire is losing air prior to reaching near-zero air pressure (i.e., various shades of blue colors in the bead flange area can be found). A compression groove (CG) also may exist. However, CGs are formed over time (cycles), and a more rapid run-flat will not add to the CG, although it is potential evidence of chronic over-deflection. On the other hand, heat evidence can form fairly rapidly. This evidence will appear only if the tire is losing air (i.e., not as a sudden air loss) and is being driven fast enough to build up excessive amounts of heat.

Without enough air pressure and with load, it does not take much time at even moderate speeds to build up enough heat to leave evidence on the tire.

6.2 Chemical Damage to the Tread and Sidewall

In discussions concerning "chemical attack" of the rubber in tires, the meaning is denoted for petroleum-type substances and products that may contain those substances. Petroleum products generally will leave an odor, can cause blistering, will significantly decrease the durometer value of the rubber, and/or will make the rubber have a spongy feel or appearance (Fig. 6.8). Due to the nature of vehicles, these products are almost always gasoline, diesel fuel, kerosene, grease, or oil.

Fig. 6.8 Chemical damage post-tread belt detachment.

Some products containing petroleum distillates that enhance the blackness of the tire sidewall accomplish this by removing the outer layer of wax (e.g., an anti-degradent added to the sidewall for ozone protection) on the sidewall surface. Although the anti-degradent will replace the lost protection by a process called diffusion [Ref. 6.1, p. 3], this reduces the reservoir of the protection in the sidewall. Over time, with sufficient bleeding out and removal of the wax, the sidewall can develop premature ozone cracking.

6.3 Non-Ozone-Related Cracking, Indentations, and Bulges

6.3.1 Cracking

Cracking in a tire occurs at stress points, either those from the internal tire structure or from the external environment. Non-flow crack "cracking" in a circumferential line along a defined radius can be caused by a design feature. The following are a few examples:

- In the lower one-third of the sidewall, cracking can occur at an internal construction ending. This ending could be a ply ending, chipper ending, or rubber ending, such as the bead filler.

- On a white sidewall tire, cracking in the upper sidewall, and sometimes in the lower sidewall near the bead centering ring, usually is the ending of the whitewall cover strip.

- In the buttress, circumferential cracking can occur from the sidewall wing-tip ending in tires with a construction feature known as tread over sidewall (TOS). (See Appendix B for a definition of TOS.)

- In the tread, a generally linear crack in the radial direction can be the tread splice.

- Cracking along an innerliner splice, of minimal depth, can result from many causes, including over-deflection, torque, foreign material, and so forth.

It is unlikely that any of these types of cracks alone will be the cause of a tire failure. They should be noted during the inspection as part of the process of understanding the general condition of the tire, and they should be analyzed in conjunction with the entire tire examination.

6.3.2 Indentations

Indentations in the tire sidewalls (e.g., Fig. 6.9) generally are not a concern and would be unlikely to cause a tire failure. The following are a few examples:

- Foreign material can lodge between the rim and the tire, leaving indentations in the CG area (Fig. 6.10).

- Ply splices will occur in a radial direction (Fig. 6.9) and are present from the time the tire exits the mold in the factory. They are more highly visible upon inflation, but after time they can be found, felt, and seen on the interior or exterior of the unmounted tire.

- The depth of spare tire marks[6-2] from the spare tire holder (Fig. 6.11) is both time dependent and dependent on the force that is used to hold the spare tire to the spare tire rack.[6-3]

[6-2] Spare tire marks form on all tires and are neither brand nor construction dependent.

[6-3] Research did not locate a study determining how fast these marks appear in a new tire. However, from personal experience, I have seen very visible spare tire marks appearing in 17 months from an under-the-truck-bed spare tire holder.

Fig. 6.9 A ply splice in the sidewall.

Fig. 6.10 External foreign material caused indentations caught between the wheel and the tire bead.

Fig. 6.11 Spare tire marks in the sidewall, noted inside the squares.

Spare tire marks almost always occur in groups, and they may be spaced at equal points around the sidewall (Fig. 6.11), in the shape of a triangle, or as two bars. It all depends on the method used to hold the spare tire to the vehicle.

The indentations shown in Fig. 6.11 are readily visible. However, all inspections of tire sidewalls should include a review of the upper to mid-sidewall by using a hand-held fluorescent light, looking for indentations that may not be readily visible.

It is not uncommon for spare tires to be substantially underinflated at the time of use, and the finding of spare tire marks may be important to the final conclusions. The spare tire shown in Fig. 6.11 is a size LT235/85R16/E and is placed on the right rear position of a 15-passenger van. The inflation pressure was only 2.3 bar (33 psi), with a vehicle placard of 5.5 bar (80 psi).

Both sidewalls should be examined for the presence of spare tire marks. It is rare, but not uncommon, for tires to have been flipped on the wheel and placed back in the spare tire holder between on-the-ground use, thereby imprinting both sidewalls with spare tire marks.

6.3.3 Bulges

Bulges in a tire may be involved in a tire failure and should always be investigated in detail. With proper lighting and a good tactile examination, some bulges can be found or at least suspected. The suspected area can be further investigated by x-ray, looking for broken ply or belt cords or any foreign material. Shearography also can be used to find separations. However, some bulges become visible only after inflating the tire on a rim. Figure 6.12 shows a bulge that was created by broken ply cords and was seen only after mounting and inflating the tire. Note the tread-to-bead-scuff abrasion in the buttress. This was the only evidence of a possible sidewall pinch, as the tire had no tactile or visual indications of broken polyester ply cords.

Fig. 6.12 Broken ply cords, appearing as a bulge.

The only benign bulges in tires tend to be rubber component splices made during the tire building process. For example, a sidewall splice bulge will not be well defined, unlike the broken ply cord(s) bulge shown in Fig. 6.12. Rather, it will be a non-dramatic thickening of the sidewall. The expert can tell nondestructively if this is truly a sidewall bulge only by taking an x-ray to eliminate other possibilities such as broken ply cords. Destructively, the sidewall splice can be found easily by cutting a radial section from the tire (bead to bead), including the entire bulge, and then cutting a cross section through the bulge. The sidewall compound can be seen easily at this point as a lapped splice.

6.4 Identification of Innerliner Conditions

6.4.1 Appearance of a Lap Splice *(Fig. 6.13)*

The innerliner splice can be identified by several methods:

- A radial[6-4] raised area extending at times from one bead toe to the other bead toe.

Fig. 6.13 A lap splice with an opening of less than 1 mm (0.04 in.) and not open to the ply cords.

[6-4] Some splices are on a slight angle and are not radial in direction.

Identification of Various Tire Conditions 131

- An excessive amount of bead toe rubber for a short circumferential distance of approximately 12 to 25 mm (1/2 to 1 in.) in the bead toe.

- A visible line, such as someone drawing a pencil line.

- A distinctive roughness, smoothness, or color difference of several inches wide versus the remainder of the innerliner appearance. This is evidence that a splice protector strip was used during the building of the tire.

6.4.2 Appearance of a Butt Splice

A butt-spliced innerliner can be difficult to identify because the radial or angled butt-spliced liner will have only a small amount of overlapped rubber. However, it may have a splice protector appearance in the area of the splice, as shown in Figs. 6.14 and 6.15. When (or if) the protector is removed after curing, this area will be smoother or rougher in texture, or a color differentiation will exist versus the surrounding liner.

Fig. 6.14 A butt-spliced innerliner splice protector.

6.4.3 Liner Tags

Identification tags (Fig. 6.15) are placed adjacent to the liner splice when the tire is removed from the drum by the first-stage builder.[6-5] Using this method to find the splice is not foolproof because the tags may have fallen off the liner or may have been placed 6 or more inches away from the splice.

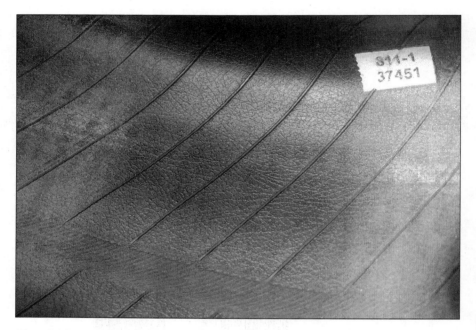

Fig. 6.15 A butt-spliced innerliner splice, with the liner tag.

6.4.4 Liner Openings

6.4.4.1 At Splice Locations

An opening of the innerliner splice (Fig. 6.16) can be formed for several reasons, including over-deflection, road hazard impacts, excessive torque, poor storage, vehicle mechanical conditions, poor building splice techniques, or foreign material.

[6-5] This assumes standard classical building methods, not automatic equipment.

Identification of Various Tire Conditions 133

An opening of the liner splice is not detrimental to the tire, unless it is excessively open, and ply cords (Fig. 6.16) or ply cord shadowing[6-6] are visible (allowing ICP).

Fig. 6.16 A butt splice open, with ply cord visible.

If the liner splice uses a butt method of liner joining, and if an opening exists through the entire liner gauge (not only open at the visible surface), then there is a breach of the liner. A lapped innerliner splice generally requires the opening to be quite wide (i.e., the width of the lap) prior to reaching the ply cords.

Generally, the ply compound is a different color of black or gray than the liner, and this can be used to determine if the opening is all the way to the ply cords. Figure 1.2 in Chapter 1 shows the color (gray tones) and texture (from buffing) differences between the liner and the ply.

[6-6] Cord shadowing (CS) occurs when you can see the "shadow" of the individual ply cords, which are still coated by the rubber matrix. Cord through liner (CTL) is a condition whereby the cord or cord twist is clearly visible. It takes experience in design, testing, or quality to make judgments on the degree of cord shadowing and to state whether or not it is detrimental to the tire. See Section 6.4.5 for more details about ply cord shadowing.

6.4.4.2 Non-Splice Locations

Innerliner cracks or splits at other locations than the ply splice do occur. They are almost always in the shoulder area of the tire, follow the ply line, and can result in a sudden air loss or excessive ICP and belt separation. On rare occasions, liner cracks in the shoulder area of the tire occur that are "off angle" and do not follow the ply line (Fig. 6.7). Both of these types of liner openings can be spotted by rolling the tire, with some force, on one shoulder or by using a mechanical tire spreader that allows the tire to be spun slowly. Good internal lighting is required when observing the interior (liner side) of the tire.

Prior to progressing from the liner through the casing and sidewall compound and into a sudden air loss, liner openings allow ICP to develop, which degrades the belt package. Over-deflection and/or excessive torque, camber, or improper tire storage are the most likely causes for liner cracks. If the prior items are ruled out, then investigation into the green (uncured) ply, liner, and sidewall component gauges as they lay up in the shoulder of the tire should be completed. Improper gauge distribution in this part of the tire can cause liner splits that follow the ply line (i.e., not angled splits). This analysis will necessitate cutting the tire, and from experience in tire design and the cured tire specifications, a determination of the appropriateness of the gauge distribution can be made.

6.4.5 Ply Cord Shadowing in the Liner

Cord shadowing is the visible appearance of the ply cords through the innerliner. Judging the nuances of acceptable levels of cord shadowing takes expert knowledge in the building, design, and testing of tires.

Depending on the type of ply cord and spacing, some cord shadowing is benign and is not a concern. However, when the ply cord or ply cord twist is visible through the liner, excessive ICP is ongoing.

The cord shadowing shown in Fig. 6.17 is visible and should be noted in an inspection. However, the depth of this shadowing is not at the point where ICP from this condition would be a given.

Identification of Various Tire Conditions

Fig. 6.17 Cord shadowing.

CHAPTER 7

Identification and Significance of Balance Weight Marks

Balance weight marks are those imprints formed into the rubber in the bead flange area of the tire, which are caused by the wheel weight clips. Keeping the phrase "balance weight marks" for the rubber imprints on the tire and "wheel weight marks" for the marks on the wheel keeps these two types of marks separate and distinct, although they are caused by the same item. In this manner, it is known which item (i.e., the tire or the wheel) is being discussed.

Balance weight marks and their use in describing over-deflection evidence have already been discussed in Chapter 4 (see Section 4.2 on over-deflection). However, balance weight marks also can be used as time-dependent events, as follows:

- **Overlapping balance weight marks with the compression groove (CG)**—Balance weight marks normally overlap in the same location as the CG. The time to formation of the CG versus the formation of the balance weight marks can give the expert a time sequence of events. In Fig. 7.1, the balance weight mark shown overrides the CG, and the base of the CG is visible just under the balance weight mark. Therefore, the CG and the over-deflection existed prior to the wheel weight clip being put at this location, and the wheel weight has been in this location for an extended time. The same is true in Fig. 7.2, but the balance weight mark has just begun the process of overriding the CG.

  Conversely, a CG that runs through the balance weight mark, as does the right balance weight mark shown in Fig. 7.3, indicates that the wheel weight

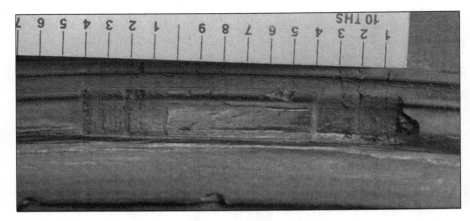

Fig. 7.1 A 2.5-cm (1-in.) wide balance weight mark, with a 5-mm (0.2-in.) shift.

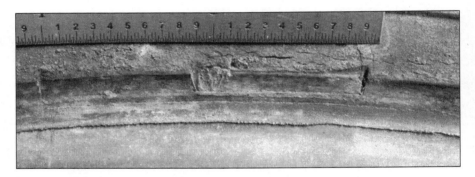

Fig. 7.2 A 2.5-cm (1-in.) balance weight mark, with the compression groove in place prior to the weight mark.

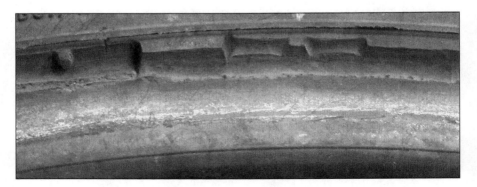

Fig. 7.3 A balance weight mark overriding the compression groove.

Identification and Significance of Balance Weight Marks

mark came before the CG. In addition, if the same types of balance weight marks[7-1] are seen but are different in depth, then the lightest mark has been in that location the least amount of time, or the over-deflection of the tire has changed. There are a number of scenarios among the CGs, the balance weight marks, and other marks that may be located in the bead area of the tire. The expert should look at all these marks as potential time-sequential events that could be important as discovery continues in the case.

- **Multiple mounting evidence**—If the tire has more balance weight marks than the wheel has wheel weights plus pre-existing clip marks,[7-2] then the tire possibly has been on another wheel. If the opposite occurs, and the wheel in front of the expert has more wheel weights than there are corresponding balance weights, then the tire and wheel were not mounted together or have been mounted together for such a brief period of time that the missing balance weight imprints do not exist.

- **Hard braking or impact shift or accelerating, usually in combination with low internal air pressure**—This type of damage can appear as a shift or chattering of the balance weight mark. Generally, the tire is moving on the wheel, but the wheel weight does not move.[7-3] Unlike a rebalance where the wheel weight is removed and replaced, leaving clean balance weight marks, shifting or chattering wheel weights will have the appearance of a wide wheel weight, with evidence of the original size embedded in the movement.

A version of the shift in balance weight marks is called "chattering," as shown in Fig. 7.4, where the chattering from the initial balance weight location (left side of the photograph) to its final location was more than 30 cm (12 in.).

Balance weight marks that chatter will leave impressions that will not be that deep or can tear some rubber. Furthermore, as the name implies, there will be multiple generally overlapping marks. This may or may not happen during the accident sequence. If it does not occur during the accident sequence, then the impressions left in the rubber will tend to be rounded due to the continued pressure of the wheel against the tire.

7-1 Different wheel weight clips leave different marks, including the width and appearance.

7-2 Pre-existing wheel weight marks are those marks left behind by previously placed wheel weights that do not exist on the wheel at the time of inspection.

7-3 The opposite side bead area should be checked for wheel weight shifts and chattering because half of the tire and wheel combination cannot move unless the other half moves, too.

Fig. 7.4 Balance weight mark chattering.

- **Multiple rebalancing**—When multiple balance weight marks exist that are not overlapping but are close enough that the lead weights on the wheel clip would overlap (Fig. 7.3) or the marks themselves overlap (Fig. 7.5), then the tire has been rebalanced. Rebalancing a tire for vibration is common in the tire retail industry; however, at times, rebalancing can mask a separation in the tire that is causing the vibration.

Fig. 7.5 A rebalanced tire, with overlapping balance weight marks.

The balance weight marks with the current wheel weights that exist on the subject wheel and those that "pre-existed"[7-4] the current inspection should be matched

---

[7-4] Pre-existing wheel weights would be those that were on the wheel prior to the inspection. These must be located and identified. Some older wheels may have multiple wheel weight marks; therefore, only the new ones should be catalogued.

for the tire and wheel location. These pre-existing wheel weight marks (Figs. 7.6 and 7.7) may be from other tire(s) than the subject tire.

Fig. 7.6 A pre-existing wheel weight clip mark, with a new clip gouge on the wheel flange.

Fig. 7.7 Pre-existing wheel weight marks on the exterior of the wheel flange.

Identification of the marks that these weights made on the wheel (Figs. 7.6 and 7.7) and their possible match to the subject tire is the primary evidence to determine the tire and wheel matching points. When the tire and wheel can be matched to each other in the exact location that existed prior to the accident sequence, then the damage can be related between the two items. Sometimes wheel weights and balance weight marks do not exist. Without balance weight marks, it is necessary to examine the wheel flange and the tire thoroughly, looking for anomalies in the wheel that could have transferred to the tire. For example, in steel wheels, the butt weld of the wheel usually is 180 degrees from the valve and can leave an identifying mark on the tire (Figs 7.8 and 7.9).

Fig. 7.8 A butt weld on a steel wheel flange.

In alloy wheels, you must search for anomalies in the alloy, such as dings, pits, or radial serrations. There also is the possibility that initial accident damage, prior to debeading, in the wheel can be lined up with that of the tire sidewalls.

Fig. 7.9 *A butt weld impression in the tire from Fig. 7.8.*

For example, when a tire hits a curb on the side of the roadway during a spin, it is possible that the wheel damage and tire scuff from the curb can be located together.

If no locating point can be found on the subject tire and wheel combination, photographic evidence prior to the accident is needed. Photographs taken at the accident scene can be helpful in tire and wheel matching only when the tire remains mounted and inflated (although the outboard facing sidewall can be determined).

As shown in Fig. 7.10, inflated tires, examined post accident, can move relative to their original position on the rim. Note that the clip mark on the wheel is at a different location than the balance weight mark on the tire. If one bead of the

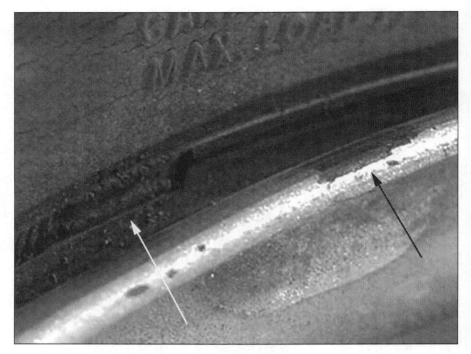

Fig. 7.10 A balance weight shift with the tire mounted to the wheel.

subject tire has become unbuttoned from the wheel during the accident sequence, you must assume that the tire and wheel may have rotated relative to one another, regardless of the discovery evidence.

Sometimes, due to either accident damage or very clean interfaces between the tire and the wheel and/or a complete lack of photographic evidence, no matching point can be found.

CHAPTER 8

Location of the Tire on a Vehicle

Locating the exact position of the tire on the vehicle by the tire inspection generally is not crucial in the long run because discovery evidence eventually will reach the expert with the location already identified. However, it might be important to identify any previous positions the tire may have occupied or in case the vehicle and/or companion tires have been destroyed.

8.1 Outboard Side Versus Inboard Side

Several methods of determining the outboard side of the tire are as follows:

- **The side with the least amount of dirt and grime usually is the outboard side**—No one washes the interior of a tire, and eventually that dirt and grime remain, even through rain and snowstorms.

- **The side with the greatest amount of ozone deterioration usually is the outboard side**—It may be only by a number or two, but the inside of the tire will be subjected to less sunshine and therefore less ozone deterioration over time.

- **Scuffing of the upper sidewall and buttress from curbing**—(Fig. 8.1) Scrubbing and scuffing the outboard sidewall against curbs (sometimes called curbing) can be prominent on the outboard side and nearly non-existent on the inboard side. Off-road use can create scuffing on the interior sidewall, but usually there will be additional abrasions in the buttress and tread shoulder. Curbing a sidewall or running up against any fixed object with a rolling tire

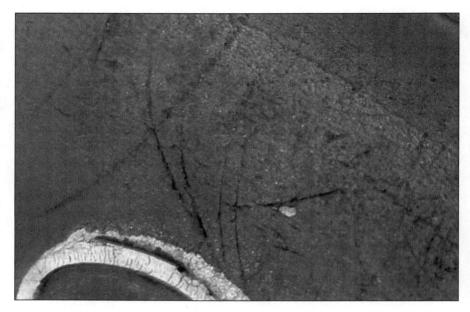

Fig. 8.1 Circular abrasions from scuffing.

will create abrasions that are curved (Fig. 8.1). Sidewall curbing can create tearing at the tops (toward the tread) of raised letters or more severe tearing that is easily recognizable as sidewall curbing (Fig. 8.2).

Fig. 8.2 Easily recognizable sidewall scuffing. All whitewall tires can scuff, as shown in this photograph.

- **Photographic evidence at the accident scene or in the storage yard**—This evidence will yield the outboard side.

- **Matching the tire to the wheel**—Such matching of the tire to the wheel will determine the outboard side. This is true during the last mounting of the tire to the wheel.

Additional comments concerning identification of the outboard side versus the inboard side of a tire are as follows:

- If it is difficult to identify the outboard side of the tire (this is more rare but not uncommon), then both sides of the tire may have been on the outboard side at some time in the history of that tire.

- Tires that have a sidewall curbing appearance on the interior are tires that have been demounted and flipped over or have been in off-road applications. Check for matching old balance weight marks and wheel weight marks.

8.2 Rear Position Versus Front Position

8.2.1 Rear Position

Several methods may be used to determine that the tire was in place on the rear of the vehicle, as follows:

- **Center wear**—On most rear-wheel-drive vehicles and especially on pickup trucks, the rear tires will tend to wear "center first." Center-first wear (Fig. 8.3) occurs when the center one-third[8-1] of the tire wears at a faster rate than the outer two-thirds. This type of wear pattern takes time to develop and will not be prominent in tires that have been rotated. However, rotation after center wear takes hold also takes time to dissipate in the front position. Using this information can determine not only that the tires have been rotated but also a general timing of when, as "not long ago" or "awhile ago." Unless additional information is obtained, estimates of the number of miles at which a rotation occurred in the past, even knowing the wear rate, would be a best guess.

- **Determination by camber wear**—Some rear-wheel-drive cars have significant camber built into the rear axle. This type of built-in camber will wear the inboard or outboard (usually the inboard) portion of the tread faster than

---

[8-1] One-third wear is an approximation. "Center wear" means the center of the tire wears at a faster rate than the shoulders. "Shoulder wear" has the opposite meaning.

Fig. 8.3 Center wear on a rear-wheel-drive vehicle.

its opposite side. Camber wear can mimic toe-in or toe seen on the fronts; however, rear-position camber wear on a rear-wheel-drive car generally will be a smoother-appearing wear rather than toe-in or toe-out because the tires are being driven in a straight-ahead fashion. Toe-in or toe-out will tend to want to feather the tread block and groove edges.

- **The rear tires on front-wheel drive vehicles will wear at a substantially slower rate than the front tires**—In this situation, the rear tires tend to form an irregular-wear saw-toothed pattern called "heel and toe" (Fig. 8.4). This pattern typically is visible across the tread surface. Heel and toe also may be visible on the front tires of front-wheel-drive vehicles, but it generally will be contained to the outer tread block elements.

8.2.2 Front Position

On the front position of vehicles, both rear-wheel and front-wheel drive, alignment and mechanical problems with the vehicle can affect the wear pattern and can cause several kinds of irregular wear (e.g., feather edge wear, scalloping wear, diagonal wear) [Chapter 4, Ref. 4.5, pp. 14–39]. However, front-position wear on all vehicles typically will yield a more balanced wear than the rears, with the shoulders wearing faster than the center.[8-2] It is not unusual for front tires to have one shoulder wearing faster than the other shoulder. The expert's background will determine if the difference in wear rate or wear pattern makes it unusual or notable.

8.3 Left Side Versus Right Side of the Vehicle

The identification of the left side position versus the right side position on the vehicle, without pictorial evidence, is the most difficult determination in all the position placements on the vehicle. To determine the side, you must determine the direction of rotation. If the tread is available for inspection, the heel and toe pattern (Fig. 8.4) can direct you. With heel and toe, the high side of the saw-toothed pattern will be toward the front of the vehicle. If the outboard side has been identified, the high side of the heel and toe then will identify the left or right side of the vehicle.

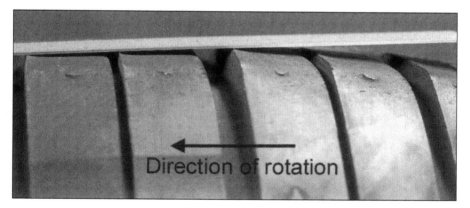

Fig. 8.4 Heel and toe wear.

With a tread and belt detachment, the casing may give some clues as to the direction of rotation if the tire has developed a clear leading and trailing edge flap and characteristic thin or thick belt skim delineations [Chapter 2, Ref. 2.1, p. 6].

[8-2] Over-deflection will exaggerate this condition.

Determination of the side of the vehicle on which the tire was located is made somewhat difficult due to possible rotations (especially those involving "X" patterns), mechanically induced wear from the vehicle, and damage to the tire. This job is made somewhat easier if you have in front of you all four tires from the vehicle.

CHAPTER 9

Addressing Several Failure Theories

9.1 Brassy Wire Failure[9-1]

Belt wire steel filaments are laminated with brass.[9-2] During vulcanization, a portion of the brass combines with the chemicals in the belt compound to form a copper sulfide (CuS) and zinc sulfide (ZnS) bonding layer.[9-3] Not all the brass is consumed in this process, leaving microscopic layers that go from bright steel (the wire itself) to brass (the laminate), to the combined rubber and brass bonding area, and finally to the rubber itself.[9-4] The fact that the brass still exists within this total bonding layer means that when the tread and top belt become detached from the bottom belt and the fracture line is within the sulfide layer, a brassy wire area may become visible in the tire.

The appearance of brassy wire anywhere in a tread and belt detachment has been construed by some experts to mean that there was no or only borderline bonding between the rubber and the belt wire.

I will cover this topic in the following three subsections.

9-1 General reference for Section 9.1 is made to Ref. 9.1.

9-2 The brass thickness is generally between 0.2 and 0.3 microns (0.000012 in.).

9-3 Brass coating and rubber adhesion to the coating of tires have been studied and written about extensively as far back as the 1970s [Ref. 9.2, pp. 604–675].

9-4 The laminate layering is shown in Ref. 9.3, pp. 8 and 13.

9.1.1 No Bonding Between the Brass Laminate and the Belt Compound

If no bonding between the belt wire and the belt compound was built into a tire, the result would be a distorted tread area on exit from the curing mold within the manufacturing facility. A distorted tread area has the appearance of a twisted tread and is easily visible. Because all tire manufacturers (in my experience) use 100% visual and tactile inspection after vulcanization, as well as 100% mechanical inspection of some type, a distorted tread would be identified within the tire factory. On the extremely remote possibility that the tire were to make it to the stream of commerce, be balanced at a retail shop, and be placed on a customer's car, distorted tread would occur within a short time and with few miles on the tire.

9.1.2 Partial Bonding Between the Brass Laminate and the Belt Compound

Partial bonding that can withstand the rigors of visual and mechanical checks at a tire factory and have the proper appearance during the retail dynamic balance and placement of the tire onto the vehicle is very rare but may occur. Partial bonding may happen, for instance, when the cure curve for the vulcanizing press does not match that of the rubber compound. The partial bonding will break down rather quickly on the road, and a distorted tread with loose belt wires within the rubber matrix will appear within a few thousand miles or several months on the road. This may result in a brassy wire appearance, or, more likely, the brass wears off, and bright steel remains. To say that partial bonding will yield many thousands of good miles and years of use before coming apart is not to understand the mechanics of this type of failure.

To have a failure of the brass-to-rubber bonding layer in a belt cable, the entire circumference of the belt cable will be loose within the rubber matrix, and a distorted tread will show up early in the tread life of the tire. Belt detachments where the #2 belt is tight to the tread and the #1 belt is tight to the casing cannot be claimed to be a brassy wire failure. Within the confines of a tire factory, it is impossible to have poor adhesion consistently only on the extreme top or extreme bottom of a circular wire. If that exists, the poor adhesion will affect the entire circumference of the cable.

9.1.3 Proper Bonding and Brassy Wire Appearance

Some experts interpret any brassy wire as a defective bond simply because a brassy wire appearance is rare but not unusual in tire failures,[9-5] and these experts know of the ASTM "H block pull-out test." The ASTM "H block pull-out test"[9-6] is a standard test in which the wires are encased in the belt compound, and then each wire is pulled individually out of the matrix. The degree of rubber tearing and coverage remaining on the brass cord on this pull-out test then is judged to a known scale. Because this test produces or should produce cohesive rubber tears, some experts insist that rubber-to-rubber tears are all that are supposed to happen.

Remembering that the brass layer still exists in the bonding area, to have brass appear in a tread and belt detachment means that on a microscopic level, the existing sulfide interface fractured before the rubber or the rubber-bonded interface fractured. Therefore, the bonded area to the rubber is intact. Because the H block adhesion test should produce rubber tears, we need to look at what conditions during a belt separation, tread belt detachment, or accident sequence could cause the sulfide/brass layer to fracture before the rubber.

This circumstance can occurs as follows:

- The belt compound approaches the glass transition temperature (Tg). This temperature will vary by compound, but it is a cold temperature (e.g., $-34°C$ [$-30°F$]).

- The modulus of the belt compound rises significantly, and/or the sulfide layer gradually changes (e.g., physiological ozone or oxidative damaged tires).

- The belt and tread package during a tread and belt detachment occurs at a rapid rate,[9-7] mimicking the effect of approaching the Tg value. During the detachment, the wires can be pulled loose.

Because the brass coating is less than 0.5 microns thick and can be rubbed off, a separation that shows brassy wire could not have existed for much time within a loose matrix of rubber and wires rubbing against each other. Therefore, the fact

---

9-5   *Federal Register* Notice, National Highway Traffic Safety Administration Denial of Defect Petition, Petition Analysis DP02-011.

9-6   ASTM International, ASTM D2229-04e1 Standard Test Method for Adhesion Between Steel Tire Cords and Rubber.

9-7   The ASTM D 2229-04e1 is run at 50 mm per minute (2 in. per minute), while a belt and tread detachment at 60 mph can be measured in feet per second.

that brass can be visible after a tread and belt detachment means that the final detachment was the cause for the visibility of brass.

When investigating a tire failure in which brass is visible, in addition to a complete inspection to determine causation of the failure, the expert should pay particular attention to the remaining tread depth, the time in service, and the physiological age of the tire.

9.2 Manufacturing Imprints—"Liner Marks"

Colloquially called "liner marks," these marks really are in a class called manufacturing imprints. These marks might include imprints or impressions made in the belt compound, tread or belt interfaces, shoulder wedge, belt edge gum strip, sidewall, and so forth.

During the tire building process, several methods are employed to ensure that the rubber components arrive to the building area with enough tack (i.e., green adhesion) to build a tire but to not have the laminate pieces adhere together before building. For belts, overlays, sidewalls, gum strips, and innerliners, these methods may include fabric, polypropylene, or polyethylene liners. The liners separate the components when they are rolled together and usually are mechanically dimpled or laced in some fashion to reduce the area that is in direct contact with the surface of the components. This textured surface creates a reduction in surface adhesion from the component to the liner and allows the component and liner to separate during the building process. The texture of the liner can be passed on to the rubber component, which is from where the term "liner mark" is derived. In the case of heavier components such as treads or sidewalls, striations into the component can be imprinted mechanically and may appear in cured tires [Ref. 9.4, pp. 9, 15, 19–26].

During inspection of a tire failure, manufacturing imprints may be called into question by some investigators as *prima facie* evidence of poor adhesion within the components or of foreign material contamination, even when the tire has been on the road for thousands of miles and years of use.

For the following reasons, a tire expert must review the entire tire failure fully to make a determination that a manufacturing imprint is the cause of the failure, as follows:

- At times, the tearing of one belt off another yields the "appearance" of a liner pattern on one belt or more belts. Careful analysis may indicate that the pattern that appears is actually at or near the same angle as the belts. The liner

Addressing Several Failure Theories 155

pattern then is not caused by a "liner" but rather by the crossing pattern of the belt cables [Ref. 9.5, p. 11].

- The appearance of a liner pattern in the center portion of the belt during belt detachment is not unusual and can occur due to the complex radius tearing of the belt-to-belt detachment across the centerline of the tire. At this point, the belt is tearing from a thick rubber plane to a thin one and crosses what is known as the transition zone [Chapter 2, Ref. 2.1, p. 13]. This transition zone is noted in several papers and occurs in many belt-to-belt detachments, including those of new tires.

- A tire is a laminate structure. A tire is *not* like a milkshake, where everything is mixed during curing, or as Herzlich states, "Boundaries between the various material laminate layers that make up a tire do not 'disappear' or 'homogenize' during the curing process" [Chapter 3, Ref. 3.1, p. 4]. A tire is built by adding layers of materials in a sequential order. If a tire is cut and sectioned properly, each individual component is visible.[9-8] When a tire tears during a tread and belt detachment, it typically and randomly tears in a multi-planar fashion, crossing the various layers or planes. However, at times, it tears along the planes of the layers. This type of tearing can yield an appearance of the original surface of the laminate material, including manufacturing imprints. Seeing this surface does not mean it was not adhered nor bonded. The chemical and mechanical bonding did occur, and the tear along the laminate plane can sever these bonds [Ref. 9.6], especially if the operating conditions generate excessive heat.

- It is not possible to have many thousands of miles, and years of use, with a partially adhered component or one with zero adhesion. As the tread contacts the road, partially adhered components or those with no adhesion act no differently than a tire with a belt separation. At 60 mph, a partially adhered area "hits" the road approximately 12 times per second.[9-9] The centrifugal and loading forces quickly will make these components tear, bubble, bulge, or cause a accelerated wear area on the tread surface early in the life of the tire (see Section 2.3 in Chapter 2). Daws is more specific, indicating that manufacturing conditions typically show up within the first third of the life of a tire [Chapter 1, Ref. 1.2, p. 20].

- Manufacturing imprints rarely are found in the belt-separated area. However, some investigators opine that the evidence of manufacturing imprints in the tire

---

[9-8] See Figure 1.2 in Chapter 1 for a section view of a tire, with all the laminates clearly visible.

[9-9] This is an approximation for the average-size passenger car tire. It will be more for smaller tires and less for larger tires.

away from the separated area is evidence that they are likely to have existed in the separated area, too, and this was the cause for the separation. Tearing within the same plane generally does not occur for great lengths of distance in a tire before it begins to tear again in a multi-planar method.[9-10] Seeing manufacturing imprints in one part of the tire does not then automatically infer that the imprints are in other parts of the tire.

There are times when imprints should be investigated as a possible cause or contributor of a tire failure. Some of these events include the following:

- Poor building practice, for example, components poorly laid up or contaminated with foreign materials.
- Pieces of non-curable materials or irregular gauge distributions found in the tire.
- Early tread life failures or early tread life accelerated wear areas and bulges or bubbles.
- Manufacturing imprints that appear on a large scale in the tire and are adjacent to the belt separation area, including early tread life failure.

In a failed tire where the expert suspects contamination as the cause of the failure, the only method to quantify the source of the contamination is the use of laboratory and chemical testing. Unfortunately, if the area in which the suspected contamination resides was exposed to the environment and has a chain of custody, it may be difficult to separate the manufactured contamination from the road or environmental contamination.[9-11] However, if another area of the tire has a "closed off" separated area, or enclosed bubble, that area is a better source for analysis. This will entail cutting the tire in a methodical manner in a clean environment.

9.3 Nylon Overlay

"Nylon overlay" is a descriptive term. The material is nylon (as opposed to steel or polyester), and the material "overlays the belts." The terms "cap ply" and "cap plies" technically are not correct. The cap in a tire describes the top part of

---

[9-10] This is not true for the belt detachment across the transition zone, which can traverse circumferentially around the tire.

[9-11] The chemist's background in differentiating manufacturing-made chemicals versus environmental ones is quite important in this analysis.

a two-compound tread, and a ply is a reinforcing material that typically wraps both beads. A cap ply, as described, is neither of these but rather is an overlay of material laid on top of the belts at a zero-degree angle.

Nylon overlays have and continue to be called "cap plies," "tourniquets," or "safety belts." The overlay is neither a tourniquet nor a safety belt. Although those terms are descriptive, they describe something that a nylon overlay is not. Nylon overlays do not prevent crack initiation, propagation, or growth leading to belt separation. Belt separation occurs in overlay tires as in non-overlay tires. Due to the overlays running at zero degrees, when a belt separation does occur between the working steel belts, the separation may be masked for a time (e.g., a delayed accelerated wear). The growth of the belt separation and possible belt detachment will be due to the same forces and will have the same appearance and causes as a tire without an overlay.

Nylon overlay structures are listed in patent literature from the late 1960s. They have been applied to speed-rated tires in Europe for years and slowly have made their appearance in the United States as speed ratings on tires became more commonplace. In tire construction, adding an overlay of nylon is the easiest method to pass regulatory or internal tire company high-speed standards. The overlay allows for a small controlled growth in the diameter of the tire at high speed. This constraint on the growth raises the speed at which the destructive "standing wave" forms in a tire.

There are several positives and negatives to adding overlays to a tire. The positives are as follows:

- High-speed capability
- Lower-conicity force values
- Somewhat improved tread wear
- Improved handling

The negatives are as follows:

- Potentially increased rolling resistance
- Increased tire complexity
- Potentially a higher shoulder temperature
- Reduced resistance to road hazard impacts
- Stiffer ride

U.S. Department of Transportation (DOT) stamping regulations for tires require that the number of plies and the type of material in the tire that crosses the centerline must be listed on the sidewall. If a nylon overlay crosses the centerline (i.e., a full-width overlay), it will be listed on the sidewall. If the tire contains nylon overlay strips that do not cross the centerline, they will not be listed on the sidewall. Either the visual inspection of a failed tire or a cut section or the use of an x-ray can determine if overlay strips exist.

The fact that a nylon overlay exists in the tire should be noted during an inspection.

CHAPTER 10

Visual and Tactile Nondestructive Tire Investigation Techniques

10.1 Basic Inspection Process

A methodical and systematic laboratory examination of the tire off the wheel is the key to a thorough inspection. If the examination is performed while the tire is mounted to the wheel, this is only a partial inspection. Although an inspection of the tire off the wheel may be unavoidable for a first inspection, a demounted tire examination should take place prior to the issuance of an expert's final report or testimony, unless there is ample reason not to disturb the bead seat.

A field examination can yield similar results and conclusions to that of a laboratory examination. However, the lack of proper lighting, setup, and tools, as well as the time constraints of a field inspection, may yield less than satisfactory documentation, photographs, and detailed causation analysis. The best option is a laboratory examination that allows several days of examination, photography, analysis, and introspection. It is not unusual to find additional markings and additive modes of failure during a laboratory examination in the expert's own laboratory that were not found during a field examination, even when the field examination is performed at an indoor location, including another expert's laboratory.

Only in rare cases can photographs alone substitute for an inspection of the tire. Determination of the cause(s) of failure from a photograph-only non-inspection of the tire can occur only if the photographs are of sufficient quantity, quality, and detail and are provided with a tire forensic expert's notes. Even if this information is provided, the expert who is relying on another expert's material is putting his or her reputation in the hands of the individual who made the notes and took the photographs, which is a potentially risky venture.

On the other hand, photographs alone may be adequate to allow the tire expert to determine that a tire disablement did not occur, such as in a rollover with the wheels facing upward, and all the photographs show mounted and inflated tires without a tread separation. In such a case, an argument could be made that no tire failure occurred. Also in cases where the tires and the wheels or the vehicle do not exist and photographs are the only remaining evidence, the expert may try to make a determination that a tire failure occurred and what general type of failure it might be. Strict causation in this case should be avoided because no actual inspection occurred.

An inspection should include the following:

- **Photographs**—These should be in color and taken with a macro lens, if needed, to be explicitly clear about what is being photographed. There should be enough photographs to adequately explain to a jury the opinion being held years after the inspection occurred.
- **Notes**—The notes should reflect only what you see. Do not note what you do not see, unless there is variability in what the inspection shows on one side or component of the tire versus another. In most situations, if something is not there, it should not be included in your notes.

The photographs should be extensive enough with the notes to defend your opinion when an opposing expert has found something in the tire that might be contested. You then will be able to review the photographs and notes and decide what the other expert is visualizing. All tire inspections, including field inspections, should be treated as if you will not see the tire again prior to the trial.

Inspection notes should not include written opinions. The expert must record professionally what is being seen, felt, measured, and photographed in a nondestructive visual and tactile examination, rather than the thought process. The exception to this is when descriptive words are used to best explain what you are seeing. For instance, the phrase "road rash" can describe the road-worn appearance of a belt after a belt detachment, whereas the written physical description of road rash may require a paragraph or more of explanation.

As you run through the following inspection technique, and regardless of the methods of inspection note-taking and photography you finally adopt, strive to be consistent and, more importantly, thorough. Using the same method over and over will bring consistency in the product you publish.

Visual and Tactile Nondestructive Tire Investigation

The following is one of many methodical inspection methods that can be used.

10.2 Marking the Tire for Inspection

Non-reflective black masking tape can be used in approximately 1 cm (1/2 in.) long pieces, placing one piece at each clock position. Also, pre-made stickers may be used to mark the tire. I prefer to use non-reflective tape, so further references here will be made only to the tape. Place the first marker in the buttress at the "O" in the molded "DOT" (Fig. 10.1), which is located in the lower serial side (SS) of the sidewall.[10-1] This is the starting position and becomes position 12 of a clock face.[10-2] Moving clockwise, begin placing markers every 30 degrees as positions 1, 2, 3, and so forth,[10-3] until returning to position 12 (Fig. 10.2).

Fig. 10.1 Make the "O" in "DOT" the twelve o'clock position.

Add another piece of black tape on the serial side tread shoulder. If the tire tread and/or belt(s) are detached, place the tape on the serial side portion of the

---

[10-1] See National Highway Traffic Safety Administration (NHTSA) regulation 49 CFR 574 for explanations of the U.S. Department of Transportation (DOT) code.

[10-2] Analog clock face nomenclature is more readily understood by non-engineers than using 0 to 360 degrees marked off every 30 degrees.

[10-3] It is helpful to have a cardboard cutout of the clock face already made. Using this prop, the clock face markings will be consistent on this tire and with others.

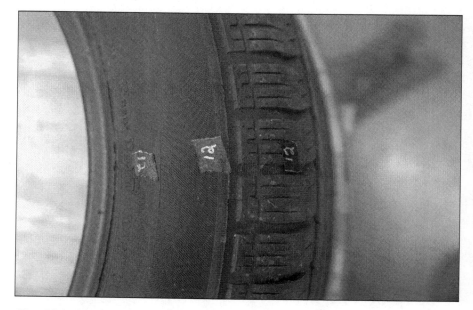

Fig. 10.2 Marking a tire for inspection with non-reflective black tape.

remaining casing. Using silver ink, add the clock face numbers onto the black tape pieces.

This method of marking provides several advantages:

- When the inspection is complete, the markers can and should be removed from the tire, thereby leaving no marks on the tire.

  Black tape does leave a small amount of adhesive that can be found by an analytical chemist. It is imperative that as few marks, tags, or writing as possible be permanently added to evidence unless there is no other choice, and approvals to mark the tire must be obtained. Permanent or semi-permanent markings obscure evidence that other experts need to see.

- In photographs of the tread or casing surface, the serial side of the tread surface will always be noted by the tape on the shoulder because no other tape is visible on the tread surface.

  For consistency, during photography the tread of a tire on a "lazy Susan" inspection stand should always have the serial side face up.

- The non-reflective black tape with silver ink shows up nicely in photographs in all but very bright conditions.

10.3 Examination Process—Notes and Photographs

1. Starting with the serial side sidewall, write down the pertinent information that is molded into the sidewall. Specifically, this includes the complete DOT, brand, tire name, size, load index, speed rating, and any other information that might be pertinent to the particular case (e.g., Economic Commission for Europe [ECE] marking, loads, pressures, or "made in" information). Although this may seem tedious because photographs will indicate all the stamping, it is important to the process of getting to know the tire. Furthermore, this basic information can be found faster than opening up a laptop and examining photographs.

2. Starting at position 12, inspect the entire serial side sidewall (including the bead area), writing into your notes all the relevant findings by clock position. Notes should list approximate depths, widths, and angles of pertinent items unless you have photographed the area with a ruler that is visible in the photograph. A portable lighting source should be used to scan the surface of each sidewall and tread. This inspection "by shadows" sometimes can help you to find small distortions and coloration variances in the sidewalls that may not be as visible with overhead lighting. For instance, spare tire marks sometimes are exceedingly light, and this method can find them.

3. Photographs of the serial side sidewall should be taken. These photographs should encompass 360 degrees of the sidewall (12 photographs), and the buttress and upper sidewall (12 photographs). Any other items of interest found during the inspection of the sidewall also should be photographed.

4. A small piece of black tape should be placed as near the serial side bead centering ring as possible at all the clock positions, while trying not to obscure any of the lower sidewall stamping. The tape then should be labeled by clock position, and the lower sidewall and bead area should be inspected.

5. Photograph the lower sidewall, bead, and compression groove (CG) area 360 degrees in the same manner as the sidewall (12 photographs) and any other items of interest found during the inspection of the sidewall.

6. While the tread face is serial side up, photograph 360 degrees (12 photographs of the tread). Further photographs will be taken after the tread has been inspected.

7. Turn over the tire with the opposite serial side (OSS) in the upward position, and repeat Steps 1 through 5 of the preceding procedure for the marking, photography, and note-taking on the opposite serial side of the tire. Do not place any tape on the tread. Note any stamping from the opposite serial side sidewall that might not appear on the serial side (e.g., UTQG) and the type of tire (e.g., whitewall [WW], blackwall [BW], or raised white outline letters [RWOL]).

8. Inspect the tread and casing surfaces, and photograph all pertinent items found. At this time, if there are detached tread or belt pieces, match them to the casing. Add non-reflective black tape with silver ink marking as to their reference position, inspect them, and photograph the top and bottom of all the pieces.

9. Inspect the interior of the tire. This will include a complete tactile examination of the liner, and note the location of any ply splice(s) and liner splice.

10. The inspection of the wheel is last. If it is necessary, the wheel can be marked with the black tape, using degrees instead of clock position and placing the tape every 30 degrees. In most cases, this will not be necessary, and a visual inspection of the wheel, with photographs of areas of interest and notes, is enough. The wheel should always be checked to ensure that it matches the tire. Locate the match points, and make a note as to the original valve location referenced to the tire clock position. In addition, a backpressure test should be performed at the maximum wheel pressure or secondarily the maximum tire pressure as listed on the sidewall.[10-4] If necessary, a bead seat diameter measurement and profile check of the wheel should be done.[10-5]

11. The final photographs that tie the package together with all the individual pictures should be of the entire tire standing up (if possible) from the opposite serial side, the serial side, and the casing or tread side. The same photograph angles should be taken with the wheel. These steps will ensure that the entire exterior and interior[10-6] surfaces of the tire and all the other pieces supplied for the inspection are photographed and inspected.

[10-4] There are exceptions to the pressure to use. Follow the Tire and Rim Association (TRA) guidelines or the vehicle placard for allowable additional pressures that can be added for extra load or speed.

[10-5] The Tire and Rim Association (TRA) in Copley, Ohio, sells these gauges and guarantees them accurate to the TRA dimensions (www.tra-us.org).

[10-6] Innerliner photographs should be taken as needed. The markers do not adhere well to the interior of the tire due to the anti-adhesion surface on the liner. If photographs are necessary, the black pieces of tape should be marked with the clock position and "SS" or "OSS" if the interior shoulders are being photographed. As an alternative, the inspector's notes should indicate which shoulder is the SS or OSS side.

Visual and Tactile Nondestructive Tire Investigation

This procedure works well in a laboratory environment, but it is time consuming and may not be possible in a field environment. Most tire inspections can and should be made with this intricacy in a laboratory environment. However, each inspection should be judged on its own merit. There will be times when the type of failure or the case dictates fewer or more photographs or notes.

10.4 Tactile and Visual Inspection of the Tire

10.4.1 Serial Side Sidewall

Beginning at position 12, flex the entire sidewall, such that the upper and lower sidewall is bent inward and outward. Any areas that may have intra-sidewall damage should be felt for softness using the forefingers and thumbs. Each item or groups of items should be noted as to location in the lower sidewall (LSW), upper sidewall (USW), or mid-sidewall (MSW).

The following is a list of several inspection notations:

- Ozone deterioration (O/D)[10-7]

- Any cracking, tearing, gouges, ply splits, coloration differences, bulges, or indentations.

- The feel of the elasticity of the tire, both inside and out. This will yield information as to the physiological condition of the tire.

- The direction of the pertinent tears (from the wheel to the tread, from the tread downward to the wheel, or in the circumferential direction) and the angle, using the radial direction as zero degrees (or 90 degrees, whichever is your preference).

- The depth of the pertinent damage, unless it is somewhat superficial.

- The side that most likely was the inboard or outboard side of the tire should be determined, if possible.

- If there is substantial or severe general damage to large parts or all of the sidewall, a general note should be made to that effect, and only those areas deemed important to the analysis should have detailed notes taken about them.

[10-7] I use the Shell rating scale that goes from 0 to 10, with 10 being a virtually new tire, and 0 being a rubber component that is completely cracked through. (See Appendix E for more details about the Shell rating scale.)

- Note anything that is not expected (i.e., anything odd or unusual, which the expert's background will determine), but do not list it as unexpected. Just describe the physical attributes of what is seen.

- Photograph each critical item about which a note is made, so that there is no question what the notes say versus the actual item on the tire. Any additional photographs taken during the examination should be taken immediately after the 360-degree photographs for the sidewalls are taken. This procedure keeps the additional photographs with the sidewall being examined in order in the camera. Then later, when viewing the photographs, it will be clear as to which sidewall the photographs belong. It is important that the photographs you take are what you see visually. This may mean trying different camera settings and lighting directions or sources.

10.4.2 Beads

- Beginning at position 12, flex each bead area 360 degrees for cracking, and feel the lower sidewall for any softness. Then each bead toe should be "lifted" up 360 degrees to feel for bead tearing on the innerliner side of the bead. Finally, a visual inspection with a portable light should be made on the interior (i.e., liner side) of the bead 360 degrees, noting any cracking or creasing by position.

- Note any bead damage by clock position and location within the bead area (e.g., bead face, bead flange area, bead toe, or liner side of the bead toe). This damage includes cracks, tears, striations, worn surfaces, coloration, hardening, lines, gouges, and so forth. As in the sidewall inspection, if there are general areas of damage, a general note can be made covering what that damage is and what positions it covers, or simply noted as "360 degrees."

- Make a note of the location of any balance weight marks, measure the width of each mark, and note the depth of each mark. There currently is no bright line standard for acceptability as to the depth. However, it is relatively easy to show photographs of deep balance weight marks versus light ones. Also, make a note about any balance weight shifting, chattering, or overlap.

- Note the depth(s), color, luminosity, scale, and general appearance of the compression groove (CG). The CG, located in the bead flange area, is an indication of chronic over-deflection. There is no bright line standard in the tire industry as to acceptability. I use the following terms to identify the degree of the CG: none, slight, slight to moderate, moderate, moderate to severe, and severe. Other experts have various methods of determining the levels of the CG, or whether or not a CG exists. It takes a lot of experience to properly judge the severity of the CG. It is dependent on time, over-deflection, tire construction, and, to some extent, aspect ratio. See Section 4.2.1 of Chapter 4 and Appendix C for further discussion of the CG.

- Note anything odd or unusual, again only describing the physical attributes of what is seen.

10.4.3 Opposite Serial Side Sidewall

Examination of the opposite serial side (OSS) sidewall is completed in the same manner as that for the serial side (SS) sidewall.

10.4.4 Tread

Buttress damage is considered part of tread damage rather than sidewall damage because it is part of the tread shoulder of the tire. The tread is inspected beginning at position 12, and all damage is noted by its general position (e.g., SS, OSS, center, or intermediate groove). As long as the tire can be raised to the vertical and manipulated easily, it is best to inspect the tire with the overhead lights shining on the tread surface. If damage, weight, or size precludes comfortably standing up the tire, the tread then is best examined laying flat on a "lazy Susan" table that is at countertop height. For consistency, with inspection of the tread in the vertical position, the serial side of the tire should be placed on the expert's left or right side for all tires inspected. This is necessary so that a "fishbone" diagram (see Appendix B, Figure B.3) made during the inspection will be consistent for tire after tire. If the tire is inspected flat, the serial side is in the upward position.

Items to be noted are as follows:

- **Whether the tire was supplied with additional tread or belt pieces**—Each supplied piece either must be noted as not belonging to the subject tire or, if possible, should note the location of where the piece mates with the subject tire. Each tread and belt piece that is part of the subject tire should be identified as Piece #X, and its location noted for photography by placing a label on both sides of the piece. Inspections where single or double strands of belt material have been supplied need not be noted or photographed unless they are critical to the determination of failure. If possible, identifying that the wire cord type is very likely the same as the tire can be helpful but is not necessary. If the wire cord type does not match the tire, this should be stated and the piece(s) set aside. For instance, the first tread piece is designated as Piece #1, its location on the subject tire is noted with black tape, and the clock position(s) is listed, matching its location versus the subject tire. Photographs then should be taken of each location of each piece and its matching location to the tire.

- **Tread depths of each primary groove and, if necessary, the shoulder slots**—These should be measured at multiple locations around the tire. (I use all 12 clock positions.) In this manner, hard evidence is available covering the tread wear across the tread face and 360 degrees around the tire. Pay particular attention not to accidentally take a measurement on any raised decks or fillets within the grooves. If you have no choice but to take a measurement on a raised deck or fillet, then you can note the depth of the raised deck line or fillet so it can be subtracted from the 1.6-mm (2/32-in.) tread wear indicators above the groove base markers to achieve the actual remaining tread depth, or note only that the measurement was taken in a non-primary goove.

- **Any type of irregularity of the tread wear or accelerated wear location(s) and size(s), plus the general condition and appearance of the tread**—This includes general cutting and chipping of the tread surface, ozone deterioration of the tread surface or grooves, and vehicle mechanically induced tread wear.

- **The durometer (hardness) of the tread**—Using a Shore A hardness instrument, measure this in several locations circumferentially around the tire and side to side. If the measurements are all similar, a simple range of values (e.g., 64 to 67) can be noted.

- **Any cuts, tears, gouges, penetrations, punctures, abrasions (including braking areas), impact marks, and so forth**—General notes covering large areas can be made rather than repeating the same note for each position.

- **Anything odd or unusual**—What is seen and considered out of the ordinary should be noted objectively.

If necessary to the failure analysis, the direction of the abrasions in the individual tread block elements should be noted. Many times, the forensics expert can identify that the tire was involved in a fishtail or in spin maneuvers by noting the differences in the side-to-side tread abrasion in the tread shoulders.

10.4.5 Belts

If the tread is detached in part or totally, the belts and casing are inspected beginning at position 12, and the information is noted in the fishbone diagram. Some of the information will be recorded with the camera, and at times, this could be a

Visual and Tactile Nondestructive Tire Investigation

better use of your time if the area being inspected would be complex to describe in words. (Indeed, a picture often is worth a thousand words.) A good photograph can always show what is being conveyed better than words. However, the camera cannot catch depths, widths, or angles easily; therefore, those need to be documented.

These are some things to note and/or photograph:

- The type of the belt tearing (e.g., #1 to #2, or #2 to #3, or any of the myriad variations), and the type of tearing (e.g., multi-planar, "stop marks").
- The beginning and ending of the separation event (assuming there was one).
- Polishing, color variations, reversion, or in general the area of belt stock degradation (BSD) of the belts and the location.
- The #1 belt lift, and the appearance of the rubber under the #1 belt.
- Penetrations, punctures, cuts, impact marks, radial carcass breaks, ply splits, belt cord or wire fractures and/or breaks, rusting, foreign materials, and so forth.
- Areas that have belt wires broken, cut, split, sliced, and so forth, and if the area is important to the tire failure analysis, the endings should be analyzed with a 15x magnifier.
- Areas where the #1 belt is detached and only casing ply cords exist.
- Anything that is out of the ordinary or unusual, again only describing the physical attributes of what is seen.

10.5 Inspection Procedure—Demounting a Tire from the Wheel

If the tire as received by the expert is mounted and is to be demounted either by the expert or under the expert's supervision, then the following sequence of inspection may be followed:

1. Inspect and photograph both sides of the tire and wheel from a far enough distance so that the entire tire is visible in the photograph. Then take close-up photographs of the wheel with only the lower sidewall visible, and then take photographs of the wheel weights with portions of the lower sidewall visible.

2. The valve location is marked on the tire in the lower sidewall with a "V" in a white or yellow crayon. There are times when the demounting of a tire from the wheel is done early in the investigative process, and marking a tire with the valve location in the "as-received" condition generally is permissible.

3. The inflation pressure is checked with a tire pressure gauge or noted to be zero, based on the appearance of the tire.

4. The wheel weight clips should be checked versus the balance weight marks on the tire to determine if there has been any movement of one relative to the other. Any movement should be noted. Because the tire is mounted to the wheel, only the top of the clip marks will be visible at the interface of the tire and wheel.

5. The demount can be videotaped for posterity.

6. A light amount of "soap" (or a non-petroleum lubricant) should be used, ensuring a smooth demount.

7. After the beads have been broken down (i.e., detached from the wheel), use the fingers to feel around both beads to the toes. Sometimes the fit is so tight that getting a hand into the wheel well is difficult to impossible. If this is the case, the next best method is a visual check, if possible. Any substantial tear should be noted and photographed prior to removing the tire from this wheel, and the tire should be marked in the lower sidewall with a short straight radial line. The same tire marking comments from paragraph 2 as previously noted apply here. The expert should watch the demount and not do the videotaping. This is necessary to keep an eye open for any rubber pieces that might have been torn previously and, with this demount, might have become detached from the bead. All detached pieces should be bagged and their location on the bead noted on the bag. At the end of the demount, the beads should be inspected immediately for any tears, and the demounting soap solution should be removed with a cloth.

10.6 Wheel Inspection

The wheel can be an essential part of the inspection process for the tire. A tire may show very little damage from contact with a road hazard, and damage in the tread area may actually wear off. However, it is possible that the wheel may have retained this damage, and matching a tire to a wheel can be telling.

To keep the inspection between the tire and the wheel separate, wheel damage is referenced to the valve stem (or the valve hole) when facing the valve (i.e., the outboard side of the wheel). Therefore, if necessary, mark the wheel flanges in 30-degree increments clockwise from the valve. While each 30 degrees is equal to one position on the clock (e.g., 90 degrees = 3 o'clock), by using degrees on the wheel and a clock face for the tire, any inspection notes immediately are clear as to which damage is where. This difference becomes helpful when describing items on the wheel in the flange area that transfer to the tire. Another difference in verbiage between the tire and wheel inspection concerning wheel weight marks is that balance weight marks are used concerning the tire, whereas existing or pre-existing wheel weight clip marks are used for the wheel. Existing wheel weights are those that exist now as the wheel is being inspected; pre-existing wheel weights are those that no longer physically exist on the wheel but have left their mark on the wheel flanges.

After the wheel is marked, note the following items:

- **The wheel stamping, including the size and U.S. Department of Transportation (DOT) stamp**—If there is a question about the integrity of the wheel, all the stamping should be noted; otherwise, only information pertinent to the tire analysis is needed. For example, even if the wheel is not in question, the fact that the wheel says "MAX 380 kpa [55 psi] MAX 862 kg [1900 lb]" might be interesting if the tire has a 448 kpa (65 psi) and 1088 kg (2400 lb) maximum load.

- **The wheel material**—Examples would be steel, alloy, or steel with chrome face plate.

- **Rubber transfer marks on the wheel**—Note these and, in particular, the flanges (Fig. 10.3), which typically are evidence of over-deflection.

Fig. 10.3 Rubber transfer marks in the rim flange area.

- **Paint worn off the flanges in steel wheels**—(see Fig. 4.45) This can be evidence of over-deflection.

- **Wearing of the alloy flanges or bead seat area**—(Fig. 10.4) Significant abrading of the alloy can be evidence of over-deflection.

Fig. 10.4 Alloy rim flange abrasion.

- **Valve identification**—The valve identification typically is located on the inboard part of the valve. Also note if the valve has a cap and/or a valve core present.

Visual and Tactile Nondestructive Tire Investigation

- **Wheel flanges**—The width and location of wheel flanges that are out of plane should be recorded. (See Chapter 4, Section 4.4 about impact.)

- **Striations, abrasions, or cuts and dents to the wheel flanges**—The direction of the striations should be noted (Fig. 10.5).

- **Any further damage**—Note any further damage that is needed for the inspection.

Fig. 10.5 Heavy striations in the rim flange.

10.7 Matching the Wheel to the Tire

Using the wheel weight clip(s) and balance weight mark(s), an attempt should be made to match the tire to the wheel. There are various reasons to check this, other than checking for correlating damage points. For example, the tire and wheel supplied may have been mixed up with another claim at the shipping point, or the tire has been with the wheel for such a short period of time that nothing from the wheel has had the time to transfer to the tire. Wheel weights can shift during accident sequences and sometime are removed from the wheel altogether.

A procedure for identifying shifting wheel weights is as follows:

- Once a wheel weight clip has been on the wheel flange for some time, grime begins to accumulate under and near the weight (Figs. 10.6 and 10.7). A shifting wheel weight then will leave the grime where it had built up, and a lack of dirt will be seen where the weight has now located. This is true for the clip and the weight.

Fig. 10.6 Grime around the wheel weight clip.

- The clip generally digs into the wheel (alloy wheels) and leaves a cut, or marks the wheel and leaves rust marks (steel) or paint removal.

Visual and Tactile Nondestructive Tire Investigation 175

Fig. 10.7 Wheel weight shift, showing dirt to the right of the weight.

- On alloy wheels, the clip, if it shifts, can leave a small circumferential groove into the wheel flange, leading us back to the starting position, as indicated in Fig. 10.8.

Fig. 10.8 Wheel weight shift, with a groove in the metal.

- On steel wheels, a groove from the clip can form into the wheel flange from a wheel weight shift. However, the exterior lead weight mark usually will indicate movement on the wheel and the original clip location.

10.8 Identifying Multiple Past Tire Balances

Past tire balances can be noted as follows:

- Older wheels that have had multiple balances on multiple tires are difficult, in that the match to the current tire becomes one of distinguishing the older wheel weight marks from the newer ones. After time, older wheel weights will have a buildup of dirt around each weight. When the weight is removed, the dirt will remain, giving an older weight mark a grimier appearance than a newer weight mark. (Most people are not meticulous about cleaning adjacent to the wheel weights.)

- Past wheel clips dig into the metal of the flange and cause a small gouge in the rim flange. If the weight is removed, over time the tire will begin to erode the crispness of this mark.

- Wheel weight clips prevent the tire from lying against the wheel flange. Over time (and faster with over-deflection), the tire will wear away the wheel flange paint in steel wheels and will wear into the alloy flange of alloy wheels. However, under the clip, the flange will be in the condition it was prior to the clip placement.

- The part of the wheel weight clip that is against the tire will have a worn appearance due to the tire rubbing against the clip. This pattern is similar in style to the worn appearance of the tire against the wheel flange. In Fig. 10.9, the alloy flange has abraded, and deposits have been made on the flange and on top of the clip. The deposits and abrasions of the clip and the flange are at the same location. Therefore, this clip and wheel can be said to have been with this tire for a long time.

Fig. 10.9 *A clip matched with the tire for a long time.*

During an accident sequence, a tire may leave transfer marks on several parts of the wheel. Note any part of the wheel where tire transfer marks exist. There are rare times, for instance, when tires arrive for inspection mounted and inflated but on demount yield evidence of rubber transfer marks across the wheel well. Obviously, this wheel, in the past, had a tire that became unbuttoned (i.e., debeaded) while in use. This may or may not be pertinent to the case at hand, but it is part of the history of the wheel.

10.9 Photography

With all deference to SAE publishing procedures, photographs of tires under inspection should be taken in color. Although a tire may seem to be black and white, the shading of various colors—from deep blues, greens, and grays—is better captured and explained with color.

A macro lens, or the ability of the camera to take very close photographs, is a must. Although some photographs are better taken at 30 cm (12 in.) or more from the subject, photographs taken very close to the surface can give excellent detail to a specific small area.

Within the last five years, digital photography is replacing 35 mm. At this point in time, 35 mm will give you a crisper photograph and the biggest possible blow-up for exhibits. However, unless you are a professional photographer, it is significantly easier to ensure that what you see in the tire can be seen in the photograph with digital cameras. In this day and age, any photograph can be doctored, including 35-mm photographs. Therefore, part of the original reason for using only 35-mm photographs finally is passé. Obviously, no one should ever attempt to revise a photograph. In a laboratory environment, the need to revise a photograph (e.g., darken, lighten, crop) should not exist. In the field, much more work may be required to obtain a good photograph. If this is the case, a request should be made to counsel to eventually have the tire sent to a laboratory (or your laboratory) for a better and more detailed inspection.

APPENDIX A

References

1.1 Gent, A.N. and Walter, J.D. (eds.), "The Pneumatic Tire," National Highway Traffic Safety Administration, U.S. Department of Transportation, Washington, D.C., 2005.

1.2 Daws, John W., "Forensic Analysis in Tire Tread Separations," *Rubber & Plastics News*, March 5, 2007.

1.3 National Highway Traffic Safety Administration Office of Defect Investigation (ODI) Report Engineering Analysis Report and Initial Decision: EA00-023, October 2001, National Highway Traffic Safety Administration, U.S. Department of Transportation, Washington, D.C.

1.4 Govingee, Sanjay, "Firestone Tire Failure Analysis Report," Firestone, Akron, OH, January 30, 2001.

1.5 "Tire Care and Safety Guide," Rubber Manufacturers Association, Washington, D.C., 2001, www.rma.org.

1.6 *2007 Yearbook*, The Tire and Rim Association, Inc., Copley, OH, 2007, www.us-tra.org.

1.7 "Puncture Repair Procedures for Automobile and Light Truck Tires," Rubber Manufacturers Association, Washington, D.C., 2005, www.rma.org.

2.1 Daws, John W., "Fractography of Tire Tread Separations," Rubber Division, American Chemical Society, Washington, D.C., April 28–30, 2003.

2.2 Brico, J.C., "Abnormal Wear," International Tire Exhibition and Conference, Paper No. 20, conference sponsored by *Rubber & Plastics News,* Akron, OH, September 21–23, 2004.

2.3 Brico, J.C., "Measuring Rate of Tread Wear in Radial Tires," *Rubber & Plastics News*, July 23, 2007, pp. 16–20.

2.4 Daws, John W., "On the Irregular Wear Over Tread Belt Separations," International Tire Exhibition and Conference, Paper No. 21B, conference sponsored by *Rubber & Plastics News,* Akron, OH, September 16–18, 2006.

3.1 Herzlich, Harold, "Belt Misalignments and Belt/Belt Tear Patterns," International Tire Exhibition and Conference, Paper No. 29C, conference sponsored by *Rubber & Plastics News,* Akron, OH, September, 2002.

4.1 McClain, Calvin P. and DiTallo, Michael, A., "Tire Examination After Motor Vehicle Collisions," in Baker, Kenneth, *Traffic Accident Collision Investigation,* 9th Ed., Chapter 8, Northwestern University Center for Public Safety, Evanston, IL, 2001.

4.2 Standards Testing Laboratories, "Compression Grooving and Rim Flange Abrasion as Indicators of Over-Deflected Operating Conditions in Tires," Rubber Division, American Chemical Society, Washington, D.C., Paper No. 51, October 21–24, 1997.

4.3 Brico, J.C., Forensic Tire Expertise LLC, "Bead Compression Grooving: Characteristics and Influence of Tire Deflection," International Tire Exhibition and Conference, Paper No. 44, conference sponsored by *Rubber & Plastics News,* Akron, OH, September 2004.

4.4 Grant, J., Continental Tire North America, "Rim Line Grooves as an Indication of Underinflation or Overloaded Tire Operation in Radial Tires," International Tire Exhibition and Conference, Paper No. 45, conference sponsored by *Rubber & Plastics News,* Akron, OH, September 2004.

4.5 *Passenger and Light Truck Tire Conditions Manual*, Tire Industry Association, Bowie, MD, 2005, www.tireindustry.org.

4.6 Bolden, G.C., Smith, J.M., and Flood, T.R., Standard Testing Laboratories, "Impact Simulations—What Happens When a Tire/Wheel Impacts a Road Hazard," *Tire Technology International*, 2005.

4.7 Bolden, G.C., Smith, J.M., and Flood, T.R., "Impact Simulations in the Lab," *Tire Technology International*, 2001.

4.8 Bekar, I., Fatt, M.S.H., and Padovan, J., "Deformation and Fracture of Rubber Under Tensile Impact Loading," *Tire Science and Technology*, Vol. 30, No. 1, January–March 2002.

4.9 Bolden, Gary, Standard Testing Laboratories, "Structural Impact Damage Under Varying Laboratory Conditions," International Tire Exhibition and Conference, Paper No. 17B, conference sponsored by *Rubber & Plastics News,* Akron, OH, 2006. (Also in "Impact Simulations—What Happens When a Tire/Wheel Impacts a Road Hazard," *Tire Technology International*, 2006.)

4.10 White, Andrew J., *Dynamics of Tire Failure—Research Dynamics of Vehicle Tires, Vol. 3*, 1st Ed., Research Center of Motor Vehicle Research of New Hampshire, Lee, NH, 1967.

4.11 Al-Quraishi, Ali A. and Hoo Fatt, Michelle, "Dynamic Fracture of Natural Rubber," *Tire Science and Technology*, Vol. 35, No. 4, 2007, pp. 252–275.

4.12 Rancourt, J., Polymer Solutions Inc., Blacksburg, VA, and Leyden, G., Akron Development Laboratory, "Anti-Degradants in Tires. Where Are They? Part I," International Tire Exhibition and Conference, Paper No. 16B, conference sponsored by *Rubber & Plastics News,* Akron, OH, 2006.

4.13 Baranwal, Krishna, Akron Rubber Development Laboratory Inc., "A Wax Bloom Study in Natural Rubber Compounds," International Tire Exhibition and Conference, Paper No. 22C, conference sponsored by *Rubber & Plastics News,* Akron, OH, September 2002.

4.14 *The Vanderbilt Rubber Handbook*, 13th Ed., R.T. Vanderbilt Co. Inc., Norwalk, CT, 1990.

4.15 *Radial Tire Conditions Analysis Guide*, 3rd Ed., Tire Industry Association, Bowie, MD, 2004, www.tireindustry.org.

4.16 "Tire Safety—Everything Rides on It," National Highway Traffic Safety Administration, U.S. Department of Transportation, Washington, D.C., October 2001, nhtsa.dot.gov/cars/rules/TireSafety/ridesonit/tires_Index.html.

4.17 Rancourt, James and Daigle, Leigh Ann, Polymer Solutions Inc., Blacksburg, VA, "Chemical Analysis of Tire Failure Surfaces," International Tire Exhibition and Conference, Paper No. 24C, conference sponsored by *Rubber & Plastics News,* Akron, OH, 2002.

6.1 Rancourt, James, Polymer Solutions Inc., Blacksburg, VA, "Analysis of Migratory Species in Vehicle Tires," International Tire Exhibition and Conference, Paper No. 23C, conference sponsored by *Rubber & Plastics News,* Akron, OH, 2002.

9.1 Van Ooij, Wm. J., "Brassy Wire Failures," Letter, 9/1/2001.

9.2 Van Ooij, Wm., "Fundamental Aspects of Rubber Adhesion to Brass Plated Steel Tire Cords," *Rubber Chemistry and Technology,* Vol. 52, No. 3, July–August 1979.

9.3 Yamauchi, Michael, Shimizu, Toshi, and Doi, Mitch, of Tokusen; and Yasunaga David, Okumura, Kazuo, and Nakayama, Takenori, of Kobelco; "Examination of Rubber-Brass Inter-Reacted Layer of Steel Cord by Cross Sectional TEM Observation," International Tire Exhibition and Conference, Paper No. 5C, conference sponsored by *Rubber & Plastics News,* Akron, OH, 2002.

9.4 Brico, J.C., Forensic Tire Expertise LLC, "Process Marks in Disabled Tires," International Tire Exhibition and Conference, Paper No. 22B, conference sponsored by *Rubber & Plastics News,* Akron, OH, 2006.

9.5 Rancourt, James D., "Do Liner Patterns Affect Tire Performance," International Tire Exhibition and Conference, Paper No. 50, conference sponsored by *Rubber & Plastics News,* Akron, OH, 2004.

9.6 Bolden, Gary, "Component Interfacial Tearing Appearances," International Tire Exposition and Conference, Paper No. 51, conference sponsored by *Rubber & Plastics News,* Akron, OH, 2004.

Reference materials for further reading regarding tire chemistry, tire mechanics, and tire inspection are as follows:

Clark, Samuel K. (ed.), "Mechanics of Pneumatic Tires," National Highway Traffic Safety Administration, U.S. Department of Transportation, Washington, D.C., 1981.

Gent, A.N. and Walter, J.D. (eds.), "The Pneumatic Tire," National Highway Traffic Safety Administration, U.S. Department of Transportation, Washington, D.C., 2005.

McClain, Calvin P. and DiTallo, Michael, A., "Tire Examination After Motor Vehicle Collisions," in Baker, Kenneth, *Traffic Accident Collision Investigation,*

9th Ed., Chapter 8, Northwestern University Center for Public Safety, Evanston, IL, 2001.

The Vanderbilt Rubber Handbook, 13th Ed., R.T. Vanderbilt Co. Inc., Norwalk, CT, 1990.

APPENDIX B

Terms

Accelerated Wear—A wear pattern that develops in a tire with an underlying (localized) unbonded tread, belts, or casing, usually a belt separation.

Angularity—Refers to the damage being seen as having an angle to it. Note that when looking at the sidewalls, 90 degrees is the radial direction, and 0 degrees is the circumferential direction, which is the same direction as the tread.

Bead Bundle (*also* Bead Core)—Encompasses all the wires or cables in the bead and any additional attachments to the bead, that is, if a bead wrap is used, the wrap that is around the bead also is part of the bead bundle.

Bead Centering Ring—A raised molded portion of the tire just above the area where the wheel flange mates with the tire after mounting. This 360-degree molded raised ring of rubber is meant to be used by the tire mounting technician to obtain a quick visual check that the tire has seated around the entire rim. For the forensic expert, it is a constant-radius molded feature in the tire where measurements can be referenced.

Bead Core—*See* Bead Bundle.

Bead Face—*See* Bead Toe.

Bead Heel—*See* Bead Toe.

Bead Toe (*also* Bead Face; Bead Heel)—See Fig. 4.55 in Chapter 4, which indicates the various bead area nomenclature.

Bead Wire—The wires that make up the bead. Beads come in various types, such as creel or tape, single wound, or cable. They are coated in bronze for adhesion to the rubber. The wires also come in several diameters, the most prominent of which is 0.00146 mm (0.037 in.)

Belt—A composite matrix of a reinforcing material (steel wires in most cases, but Kevlar®, nylon, and fiberglass also have been used) and rubber. It will lie on top of the plies in the crown of a tire. The belt will lie at a different angle than the casing plies and will not be connected to the plies, except as they touch near the tire crown.

Belt Cable—The accumulation of the belt wire filaments formed into a cable; the belt cable specified by the structure of the cable and the filament sizes. For instance, 2+2x.25 is two filaments wrapped around two filaments, with all of them being 0.25 mm (0.01 in.) in diameter. *See also* Belt Wire.

Belt Coat Stock—*See* Ply Coat Stock.

Belt Compound—*See* Ply Coat Stock.

Belt Edge—When a belt is placed into a tire, each belt has two edges. The belt edges of the working belts normally are the highest stress point of the tire.

Belt Edge Gum Strips—A rubber component that spreads the edges of the working belts apart for some "x" distance. It generally is placed from the edge of a belt inward, toward the centerline. In two-belt tires, this gum strip generally has one end at the #1 belt edge and the other end inward toward the centerline of the tire.

The gauge, width, and placement of this strip can vary, depending on the tire size and brand, load range, speed rating, and so forth. Each company has its own gauge (i.e., thickness), compound type, and width for this strip. In addition to their placement between the working belts in the tire, the belt edge gum strips also can be placed around the edge of the cut belt edges, or between or on top of any additional belts.

Although most tires today use belt edge gum strip construction, it would be perfectly acceptable not to have gum strips present if the tire were designed not to have them. Two examples of this are as follows:

- A thick rubber gauge of the entire belt structure is used, which basically equates to the additional gauge of the belt edge gum strip.
- Folded belts or triangulated woven belts.

Belt Edge Region—As seen in a cross-sectional view of the tire, this is an area that is approximately 1/2 in. in diameter and is centered on the belt endings.

Belt Step-Off—(Fig. B.1) The horizontal distance from the edge of one belt to the edge of another belt. It can be used to describe the edge of the #1 belt to the edge of the #2 belt, or the #2 belt to the #3 belt, and so forth.

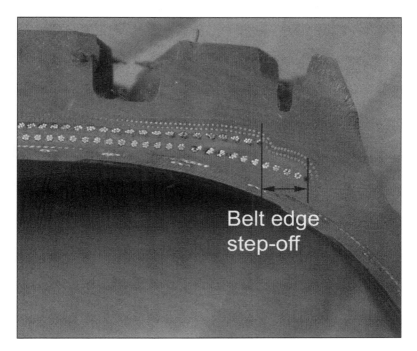

Fig. B.1 Belt step-off.

Belt Stock—*See* Ply Coat Stock.

Belt Stock Degradation (BSD)—A general phrase to describe any of the conditions seen in the belt compound, as described in Chapter 2, Section 2.1.2, usually but not always within a separation area. Therefore, this phrase covers basically anything but good tearing.

Belt Structure—Describes the belt cables, the rubber matrix surrounding the belts, the belt edge gum strip, and any belt edge treatments, including nylon overlays.

Belt Wire—Wires that are the filaments that make up a belt wire cable. *See also* Belt Cable.

Belt Wire Filaments—Individual wires that make up a belt wire cable. The number of filament wires per belt wire cable varies, depending on the belt cable configuration specified by the tire manufacturer. *See also* Belt Cable.

Belts, Working—Working belts are those that carry most of the load in a tire. Because all belts work in pairs, two belts in each tire are, by necessity, working belts.[B-1] In most passenger and light truck belts, there are two belts,[B-2] and both of these belts by definition are the working belts. In rare instances in light truck and passenger car tires, three belts have been used. In a three-belt construction, usually the #2 and #3 belts are the working belts; however, the type of belt cord, the angle, and the width might make the #1 and #2 belts the working belts.[B-3] In even rarer cases, there might be four belts in light truck tires. This type of belt structure would mimic that found on radial medium/heavy truck tires. If there are four belts, the #2 and #3 belts are the working belts. In the case of a four-belt product, the #1 belt typically becomes the transition belt between the forces generated by the casing, whereas the #4 belt generally, but not always, is a belt used to provide extra stone drilling resistance or more separation between the bottom of the tread groove and the working belts for regrooving.

Bluing—The general color that results in rubber that has been subjected to a high heat buildup in the tire. The actual colors can vary from shades of blues to purples to dark greens.

BSD—*See* Belt Stock Degradation.

Buttress—The exterior surface part of the tire between the upper sidewall and the tread surface edge.

BW—*See* RWOL.

Chafer, Fabric or Rubber—A material that surrounds the bead area from inside the bead toe to somewhere near the bead centering ring. The chafer is a material chosen to help with rim/tire abrasion in the compression groove (CG) and to help prevent bead toe tearing during the mounting or demounting process.

[B-1] There are and have been many types of belt structures used or patented over the years.

[B-2] Belt counts begin with the belt closest to the tire plies (i.e., farthest from the tread). The belt closest to the plies is the #1 belt, the next one is the #2 belt, and so forth.

[B-3] For instance, a substantially narrower #3 belt than the #1/#2 belt combination and/or a high elongation (HE) type wire, potentially with a substantially different belt angle. The tire expert's background and knowledge will determine which belts are the working belts.

Cord (*also* **Cord Twist**)—(Fig. B.2) The fabric reinforcement in the tire casing. This usually is rayon, polyester, or nylon. The cords are twisted together with two or three strands, with "x" many twists per inch.

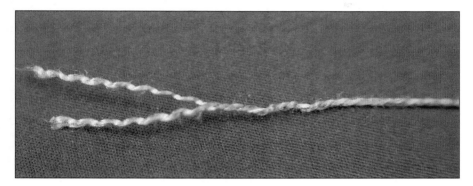

Fig. B.2 *Cord (cord twist).*

Cord Twist—*See* Cord.

Cover Strip—A thin-gauge strip of rubber that is placed over the white sidewall during the manufacturing process. It is larger in width than the white sidewall and has the ending located in the lower and upper sidewall.

Creel (*also* **Creeling**)—Steel belts typically are assembled using a method called creeling. This is different from calendaring, which occurs for fabric (e.g., polyester/nylon) and rubber materials. A creel is the machine that brings individual steel cords from individually wound spools and belt compound together, making a belt sheet.

Creeling—*See* Creel.

Debeaded—*See* Unbuttoned.

DOT—*See* U.S. Department of Transportation.

Fast Wear—Tread wear that occurs faster than expected, without an underlying belt separation. This is not accelerated wear, which has an underlying separation of some type. *See* Accelerated Wear.

Fillet—*See* Raised Deck.

Fishbone Diagram—(Fig. B.3) A method of displaying many of the findings from the tire inspection on one diagram.

Flow Crack—A line in rubber having the appearance of a crack or a tear, but with the exterior surface somewhat rounded. Probing the flow crack, if it can be opened, will show smoothness to the interior surfaces.

Green Adhesion—Uncured rubber is considered "green" or in the green state, as opposed to the vulcanized state. Rubber in the uncured state has a natural tackiness. This tackiness sometimes is referred to as "green adhesion."

High Speed—Technically, when tires are run at speeds that are near or beyond their speed capability or speed limitation. (Some tires are speed rated, and some are speed limited.) In general terms, high speed encompasses this definition, as well as highway speeds. (See Section 1.3.2 of Chapter 1 and Section 4.11.1 of Chapter 4 for more information about high speed.)

Holography—*See* Shearography.

Intact Belt and Tread Package—When there is no outward visible or tactile indication of belt separation in a tire.

Intra-Planar Tearing (*also* Planar Tearing)—Tearing of the rubber during a belt detachment, with that tearing along the laminate planes of the tire pieces or between them.

Layup—The sequence of putting the tire components on a building drum during the building phase of the tire.

Liner Cracking—Splits or openings in the liner that do not include the liner splice.

Low Profile—This definition refers to the aspect ratio of the tire but continues to change with time. Not long ago, 60 series tires were considered low profile. In today's manufacturing environment with 25 series tires, low profile generally means any aspect ratio of 50 or less.

Penetration—Objects that enter the tread, belts, or plies of a tire but do not breach the innerliner (i.e., do not enter the air chamber).

Terms

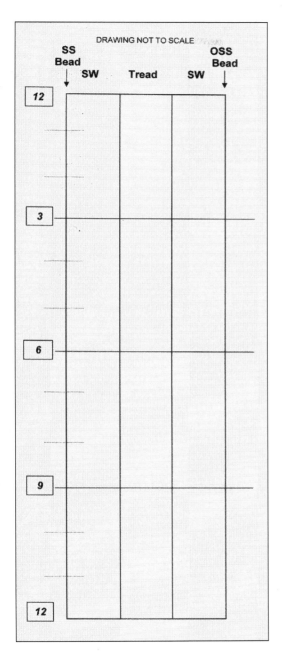

Fig. B.3 *Fishbone diagram.*

Physical Damage—*See* Physiological Damage.

Physiological Damage (*also* **Physical Damage**)—The appearance of the tire, which includes the elastic feel, durometer values, and general appearance (weather checking), apart and separate from the chronological age.

Planar Tearing—*See* Intra-Planar Tearing.

Ply Coat Stock (*also* **Belt Coat Stock; Belt Compound; Belt Stock; Ply Compound; Skim Stock**)—The rubber that surrounds the ply cords or belt cables.

Ply Compound—*See* Ply Coat Stock.

Puncture—The result of an object penetrating the tire in a defined area, from the exterior, and entering the air chamber.

Radius of Curvature—This phrase can be used to define any radius in a tire. However, in this book, it is used to define the radius of the innerliner at the tread shoulder or the tread surface juncture with the sidewall in the shoulder area of the tire.

Raised Deck (*also* **Fillet**)—In a cross-section of the tire, the deck line of a tire is a radius swung on an arc that touches the base of all the major grooves. Any tread rubber that lies above this radius (i.e., toward the tread surface) and between the tread blocks in a circumferential manner is considered a raised deck. A fillet is similar, except that it will be local in nature, connecting individual tread blocks.

RBL—*See* RWOL.

Reverse Bead Angle—Each rim of a tire has a bead seat taper angle. (Some are 0, 5, or 15 degrees.) The bead is molded to an angle from the curing hardware during vulcanization. Some tires are formed with single bead taper angles (e.g., 5 degrees), and some are dual angles (e.g., one half of the bead seat from the heel to the midpoint of the bead face is 5 degrees, and the remaining half to the toe is 15 degrees). In a reverse bead angle, the first angle is nearly unchanged from the original molded angle (matching the wheel profile at that location), whereas the second angle becomes less (at times substantially less) than the original angle (i.e., approaches 0 degrees or going negative).

Reversion—When applied to a visual inspection, this term refers to an appearance of gumminess or to a porous condition or glassiness (i.e., embrittlement) of the belt rubber. To a polymer chemist, reversion is a specific term that refers to the negative (–) slope on a rheometer curve (i.e., loss of mechanical properties).

Road Rash—Slang phrase for the abrasion seen on the top of the bottom belt that has a tread belt detachment, has remained inflated, and has been run on the road surface (or off-road) for a time.

Root Cause—Any basic underlying cause that was not in turn caused by more important underlying causes. If the cause being considered was caused by more important underlying causes, those more important underlying causes are candidates for being root causes.

RWL—*See* RWOL.

RWOL (*also* BW; RBL; RWL; WW)—RWOL stands for "raised white outline letter." BW stands for "blackwall." RBL stands for "raised black letter." RWL stands for "raised white letter" (non-outlined). "WW stands for "whitewall."

Shear—A straining action where applied forces produce a sliding or skewing type of deformation.

Shearography (*also* Holography)—Both of these methods are related techniques, using lasers and vacuum, to find anomalies nondestructively in a tire. Because they draw a vacuum to "pull apart" the separations to make the separations visible, any areas that have an open air channel to the exterior of the tire cannot have a vacuum pulled and therefore will not show an anomaly. Either of these methods can find very small anomalies. The newest versions of shearography and its software make the output much easier to read for the non-shearography expert. However, similar to all testing equipment, knowledge of the type of output and the reading of that output are best left to an expert in the field.

Shoulder Wedge—A piece of rubber that is shaped like a wedge and is put onto the casing. This rubber piece fits in the shoulder area of the tire between the casing and the bottom of the #1 belt. The #1 belt edge rests on the shoulder wedge. This piece of rubber can be a piece of rubber separate unto itself, or it can be part of the entire sidewall extrusion. Michelin found that this piece of rubber helps reduce belt separations in the tire.[B-4]

Sidewall Over Tread (SOT)—*See* Tread Over Sidewall (TOS).

Sipe (*also* Siping)—A small slit or slits within the tread blocks. These are molded into the tire at the time of cure and can be contained completely within the tread block, or may exit the tread block and join the grooves or slots. There is a term

---

[B-4] Michelin Patent No. 3717190, U.S. Patent Office.

called "micro siping," which refers to small slits cut into the tread by a tire dealer. These add-on cuts are not the siping discussed in this book.

Siping—*See* Sipe.

Skim Stock—*See* Ply Coat Stock.

Slot—A groove in a tread that can be in a general radial direction (perpendicular to the circumferential centerline of the tread) or have an angle but will not be at 0 degrees.

SOT—*See* Tread Over Sidewall (TOS).

Strain—A dimensionless engineering term for the ratio of the change in length of an object over the original length.

Stress—The ratio of force per unit area. The area is considered the original area prior to the force application.

Sub-Tread (Base)—*See* Tread Cushion.

Tears, Compression (Rubber)—The crushing of rubber or fabric, generally in the macro state, that has the appearance of a tear.

Tears, Multi-Planar (*also* Tears, Planar)—Tears that cross through the various planes of the components or tearing within a single component on multiple levels. A tire is a laminate structure, and tearing across these laminate planes or cohesive tearing on multiple levels within a single laminate is multi-planar tearing.

Tears, Planar—*See* Tears, Multi-Planar.

Tears, Tensile (Rubber)—A piece of rubber that is pulled apart.

TOS—*See* Tread Over Sidewall (TOS).

Tread Block Spreader—A tool that is used to spread the tread blocks apart. Although this specialized tool can be purchased, a large ring clip spreader purchased at a hardware store can do the same job.

Tread Cushion (*also* Sub-Tread [Base])—The tread cushion is a laminate layer applied to the bottom of the tread during the manufacturing process, usually for added green adhesion. However, this component also can be added to address ride, handling, comfort, and noise issues. A sub-tread is the bottom part of a multi-piece tread. A sub-tread contains more rubber than a cushion would contain and usually is a lower-heat-generating component than the main tread.

Tread Over Sidewall (TOS) (*also* **Sidewall Over Tread [SOT]**)—A type of tire building method in which the tread is placed on the casing after the sidewalls have been applied. The tread is stitched "over the sidewall," giving it that name. Sidewall over tread (SOT) construction can be made by several methods, but the final stitching is the sidewall over the tread. These sidewall and tread endings can become visible in the tread buttress or in a section view of the tire.

Tread Shoulder—An area of the tire where the tread edge meets the sidewall.

Turn-Down—*See* Turn-Up.

Turn-Up (*also* **Turn-Down**)—This phrase refers to the location of the ply endings around the bead, from the viewpoint of a cross section taken from the tire. With turn-ups, the endings of the plies lie on the exterior of the bead bundle. Turn-down plies lie on the interior side of the bead bundle, or they do not go around the bead at all but lie on the outside of the bundle within 25 cm (1 in.) or so of the top of the bead bundle.

Unbuttoned (*also* **Debeaded**)—The position of the tire versus its position on the rim bead seat. A tire that is unbuttoned from the rim is physically on the rim, but the bead or beads are no longer seated; therefore, the tire cannot hold air. A tire that is buttoned on the rim can hold air because the beads are seated properly.

U.S. Department of Transportation (DOT)—This U.S. government entity has a set of regulations that manufacturers must meet in order to be permitted to sell tires in the United States. These regulations are listed in the Federal Motor Vehicle Safety Standards (FMVSS) with the National Highway Traffic Safety Administration (NHTSA). The DOT code listed on a tire is proof that the tire meets these regulations. (See www.NHTSA.org for details on the codes and regulations.)

Wear Rate—In the United States, the wear rate of a tire is measured in miles per 32nd (inches) or miles per mil (thousandths of an inch). For instance, if a disabled tire has 64,000 km (40,000 mi.) and has 4 mm (5/32 in.) tread rubber remaining in a non-accelerated wear part of the tire, and the original tread depth was 9.5 mm (12/32 in.), then the wear rate is $64,000/(9.5 - 4.0) = 11,636$ km or $40,000/(12 - 5) = 5714$ mi/32nd. This term can be used in an accelerated wear area or a normal wear area.

Working Belts—*See* Belts, Working.

WW—*See* RWOL.

APPENDIX C

Compression Groove

The compression groove (CG) forms, under load and pressure, due to the rotation of the lower sidewall over the rim flange, which causes a bending moment between the midpoint of the loaded sidewall and the top of the bead bundle. This bending moment "rocks" the bead at the bead rim flange interface. Over time (miles), this rocking action eventually leads to rubber displacements that can be seen and evaluated. Over-deflection of the tire will enlarge the amount and type of displacements seen. This displacement of rubber is a compression set or physical movement of the rubber that comes from over-deflection of the tire over miles.

The CG appears in the form of texture, color, luminosity, creasing, rubber movement, and scaling variations from the original tire profile and rubber color. If the original profile of the CG is known, then measurements of the depth and shape of the CG can be determined accurately. Because the original profile generally is unknown, a measurement of its depth off the original could be quite inaccurate.

There is no bright line standard developed. Each tire forensic expert uses his or her background in making that determination. In my experience, for tires used in a proper manner, a slight or slight-to-moderate CG (my terminology) is acceptable during the lifetime of the tire. As discussed in Chapter 4 (Section 4.2 on over-deflection), the formation of a moderate, moderate-to-severe, or severe CG is evidence of chronic over-deflection.

Numerous publications have been written about this subject, as noted by the references listed in this book. The following examples of CGs do not cover the totality of all examples, nor are they meant to limit any expert from his or her own interpretation of the CG, depending on the circumstances of each tire inspection.

Characteristics and Definitions of the Compression Groove

Color The variation of color off the normal "grey." This color variation will be shades of purples, blues, and violets with possible tints of red. There also can be dark shades of gray.

Texture This is the roughness within the displacement area and includes scaling.

Displacement The amount of rubber that has been moved or "compression set," away from the original bead ring profile. Normal profiles in this area are smooth concave and convex radii. High displacement of rubber will tend to make the radii very small or come to a point.

Band Width Sometimes the CG will form two or more distinct bands. The displaced width of the circumferential band(s) that form from the displacement of rubber is the band width.

Rubber Creasing With continuing rocking from over-deflection, the rubber will form a slightly irregular circumferential area with the appearance of a crease.

Rubber Scaling The bead area rubber is rolled or flaked off and then is pressed again between the rim flange and the tire bead area, giving the impression of scaling.

Rubber Cracking The creasing may form circumferential cracks that may reach to or through any fabric or rubber chafers and the ply cords.

Examples of Compression Grooves

Figures C.1 and C.2 show two examples of slight CGs. Figures C.3 and C.4 show two examples of moderate CGs, and Figures C.5 and C.6 show two examples of severe CGs.

Fig. C.1 *An example of a slight CG.*

Fig. C.2 *Another example of a slight CG.*

Fig. C.3 *An example of a moderate CG.*

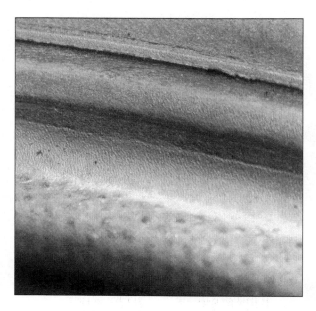

Fig. C.4 *Another example of a moderate CG.*

Fig. C.5 An example of a severe CG.

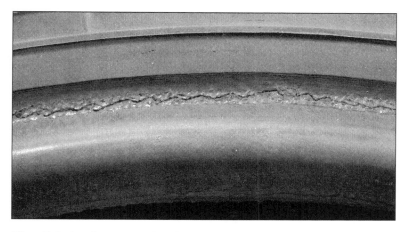

Fig. C.6 Another example of a severe CG.

APPENDIX D

Run-Flat Sequence

The following sequenced photographs are actual photographs, courtesy of Bill Haggerty of San Diego, California. They follow a car where the right rear tire was traveling without air. As the photos begin, the intact tread belt package has just severed from the sidewalls while the driver continues driving on only the two sidewalls. The license plate has been erased to protect the anonymity of the unknown driver.

Fig. D.1 *As this vehicle travels without air in the right rear tire, the intact tread belt package severs from the sidewalls.*

204   *Tire Forensic Investigation*

Fig. D.2 *The tread belt package moves behind the car as the vehicle continues down the road.*

Fig. D.3 *The tread belt package rolls across the road.*

Run-Flat Sequence 205

Fig. D.4 *The tread belt package reaches the side of the road.*

Fig. D.5 *The treat belt package comes to rest near the side of the road.*

Fig. D.6 *The vehicle has continued moving on only two sidewalls.*

Fig. D.7 *A rear view of the vehicle as it continues to move on only two sidewalls.*

Fig. D.8 Finally, the driver pulls over, without the tread belt package.

APPENDIX E

Shell Rating Scale for Ozone Deterioration

Exposure to ozone inevitably causes cracking of the rubber in tires. The Shell rating scale provides a visual means of measuring the level of such ozone deterioration.

Figure E.1 illustrates the various levels of ozone deterioration, according to the Shell rating scale. In this scale, "0" indicates the worst level of deterioration, whereas "10" indicates essentially a new tire with no ozone deterioration.

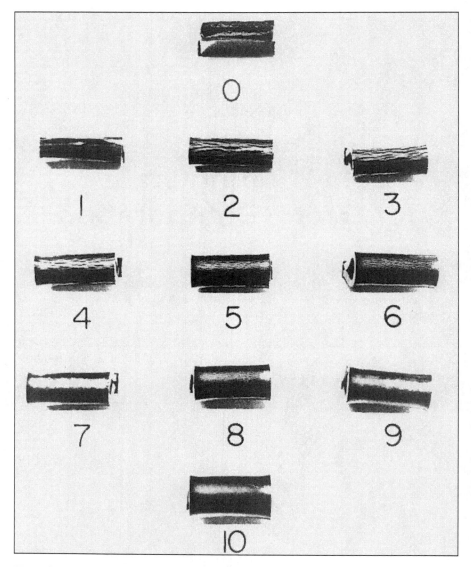

Fig. E.1 The various levels of the Shell rating scale.

Index

Abnormal wear, 110
Accelerated wear, 23, 24f, 29–33
Abrasions, 12, 21–23, 86
   bead face, 66
   circumferential, 124
   identification of, 102–104
Air leakage into the casing, 43
Alignment, 11, 122, 148
ASTM D-1149, 90n
ASTM D-2229-04e1, 153n

Balance weight clip marks, 174
   chattering of, 63–64, 139, 140f
   significance of, 137–144
Beach marks, 20–21
Bead
   breaks in, 116, 117f
   creel-type, 116
   inspection of, 166–167
   creased or cracked, 64–65
   separations of, 112–117
   tears of, 51, 93–98
Bead face
   abrasion of, 66
   circumferential line in, 66, 82
Bead wires, tensile breaks of, 117f
Belt compound, fractured, 81
Belt edge, and high temperature, 108–109

Belt edge gum strip, 3–4, 35–36
Belt edge separation (BES), 35–37
Belt package, 2, 3f
   probing for, 69
Belt separations, 1–13
   accelerated wear and, 29–33
   additive effects causing, 34
   atypical, 37–38
   causes of and contributors to, 6–13,
      39–110
      abrasions, 102–104
      cuts, 87, 102–104
      cutting and chipping (C&C),
         104–106
      demounting damage, 51, 93–98
      gouges, 72f, 102–104
      high speed, 108–109
      high temperature, 107, 109–110
      impacts, 51, 73–88
      mounting damage, 51, 93–98
      maintenance, improper, 45–46,
         107–108
      over-deflection, 56–67
      ozone deterioration, 76, 77f, 88–93,
         100–101
      punctures, 39–56
      penetrations, 67–73
      physiological aging, 98–101
      snags, 102–104
      storage, improper, 106–107

Belt separations *(continued)*
   causes of and contributors to *(continued)*
      tears, 102–104
         vehicle-related conditions, 109–110
   "clean," 78*n*
   cumulative damage and, 33–34
   edge, 35–36
      incipient, 37
   identification of, 15–34
   interfacial, 38
   localized, 81–82
   parabolic form of, 2*f*, 4, 5–6, 15
   root causes of, 6–13
   and run-flat, 121
   top belt and tread intact
      exterior characteristics, 23–26
      interior characteristics, 26–29
   tread and belt detached, 15–23
Belt skim stock, 3
Belt stock degradation (BSD), 21
Belt wires, 45
   brassy wire failure, 151–154
   breakage of, 55–56, 70, 87
      flex, 27, 29*f*, 87
      tensile, 27, 28*f*, 87
   cut, 30*f*, 87
   ozone deterioration in, 93*f*
Belts, inspection of, 168–169
Bluing, 15, 17*f*, 18*f*, 21, 56, 58*n*, 59, 66, 108*f*
Bonding, 152
Brake drag, 110
Braking, hard, 139
Brassy wire failure, 151–154
Brittle appearance, 18, 66*f*
Bubbles, 26
Bulges, 23, 25*f*
   non-ozone-related, 129–130
Butt splice, appearance of, 131

Camber wear, 147–148
Cap ply, 156–157
Center wear, 147, 148*f*
Centrifugal force, 8, 30
Chafer, 115
Chemical damage, 124–125
Chipper, 115
Chipping, and cutting, 104–106
Circumferential tearing, 50*f*, 101
Clip marks
   balance weight, 63–64, 137–144
   wheel weight, 61–62, 137
Clock face nomenclature, 161
Compression grooves (CG), 124, 126, 166–167, 197
   and balance weight clip marks, 137–139
   characteristics of, 198
   examples of, 198–201
   observation of, 56–58, 61–62
Cord shadowing (CS), 133, 134, 135*f*
Cord through liner (CTL), 133*n*
Crack growth, 1–2, 42–45
Crack initiation, 1, 3–6, 13, 42–45, 56, 105
Crack propagation, 1, 3–6, 42–45
Cracks
   circumferential, 56
   at edge, 20
   flow, 112, 113*f*
   of interior bead toe, 64–65
   non-ozone-related, 125–126
   torque, 123
   in tread, 23, 25*f*
Cross-hatching, 99, 100*f*
Curbing, 145–146, 147
Curing, 1*n*
Cuts, 12
   identification of, 87, 102–104
   from mounting/demounting, 93
   wheel flange, 98
Cutting and chipping (C&C), 104–106

Damage trail, 79
Deflection, 7*f*
  normal, 6
  over-deflection, 6–7
Demounting, damage from, 11
  identification of, 51, 93–98
Design conditions, 13
Dry rot *See* Ozone deterioration
Durometer, 98–99, 107

ECE R30, 7
Edge cracking, 20
Elasticity, 99–100
Ethylene propylene diene monomer (EPDM), 91

Federal Motor Vehicle Safety Standards (FMVSS), 2*n*
Field examination, 159
Fishbone diagram, 167
Flex wire breaks, 27, 29*f*, 87, 108
Flow cracks, 112, 113*f*
Fluorescent light, 128
Front tires, 148–149

Glass transition temperature, 153
Glossary, 185–195
Gouges, 12
  identification of, 72*f*, 102–104

H block pull-out test, 153
Hardness values, 98–99
Heel and toe, 148, 149
High speed, identification of, 108–109

High temperature, identification of, 107, 109–110
Holography, 2

Impacts, 8, 46, 51
  identification of, 73–88
  and run-flat, 122
  sharpness of fracture area, 87
  time to failure, 73, 75–76
Inboard side vs. outboard side, determining, 145–147
Incipient belt edge separation, 37
Indentations, non-ozone-related, 126–128
Inflation, 6
Innerliner, 11, 27, 41, 51, 52–53
  deformation of, 82–83
  discoloration of, 58–59
  identification of conditions of, 130–135
  identification tags on, 132
  manufacturing imprints on, 154–156
  openings in, 132–134
  ply cord shadowing in, 134, 135*f*
  radial split in, 86–87
  split in, 49*f*
  wrinkling of, 58, 59*f*
Inspection, 39*n*
  basic, 159–161
  marking for, 161–163
  note taking during, 160, 163–165
  process for, 163–165, 169–170
  by shadows, 163
  tactile, 165–169
  visual, 165–169
  wheel, 170–173
Interfacial belt separation, 38
Intra-carcass pressurization (ICP), 11, 43, 47–53, 86–87, 101, 134
Irregular wear, 31, 108

Laboratory examination, 159
Lap splice, appearance of, 130–131
Left side vs. right side, determining, 149–150
Liner *See* Innerliner
Liner marks, 154–156
Localized treadwear, 23, 24*f*
Luminosity, 18, 56

Maintenance, improper, 45–46, 107–108
Mal-design, 13
Mal-manufacturing, 13
Malwear, 10, 110
Manufacturing conditions, 13
Manufacturing imprints, 154–156
Medium/heavy truck tires, 65*n*
Modern radial tires, 1
Mounting
    damage from, 11, 51, 93–98
    multiple mountings, 139

National Highway Traffic Safety Administration (NHTSA), 98*n*, 107, 161*n*
Noise, increased, 5–6
Non-belt separations
    bead, 112–117
    sidewall, 117–118
    tread, 111–112
Nondestructive investigation techniques, 159–177
    fluorescent light, 128
    holography, 2
    identifying rebalances, 176–177
    inspection procedure, 159–161, 163–165, 169–170
    marking for inspection, 161–163
    matching wheel to tire, 174–175
    photographs, 159–160, 163–165, 166, 177
    shearography, 2, 128
    tactile inspection, 165–169
    visual inspection, 165–169
    wheel inspection, 170–173
    x-rays, 54–55, 73, 129
Nylon overlay, 156–158

Opposite serial side (OSS), 164, 167
Outboard side vs. inboard side, determining, 145–147
Over-deflection, 6–7
    identification of, 56–67
    and tire storage, 106
Overlay, nylon, 156–158
Overload, 6
Oxidation, 9, 50, 52, 86, 101
Ozone deterioration, 9, 105, 145
    identification of, 76, 77*f*, 88–93, 100–101
    Shell rating scale for, 165*n*, 209–210
    and tire storage, 106

Penetrations, 12–13, 105
    identification of, 67–73
Petroleum products, damage from, 106–107, 124–125
Photographs, 159–160, 163–165, 166, 177
Physiological damage, 9–10
    identification of, 49*f*, 51, 98–101
Ply cords, 11
    broken, 87, 129*f*
    shadowing of, 133, 134, 135*f*
    fluffing of, 46, 47*f*, 73, 74*f*
    separation between, 117–118
Polishing, 15, 16*f*, 17*f*, 18, 21
Pressure gradient, 51
Probing, 39, 41, 69
Punctures, 80*f*, 82, 105
    identification of, 39–56
    improperly repaired, 10

intra-carcass pressurization, 47–53
probing of, 39, 41
and run-flat, 122
unrepaired, 10, 40*f*
wedge-shaped, 43*f*

Radial tires, modern, 1
Rear tires, 147–148
Rebalancing, 139, 140
   identifying, 176–177
Removal from service, 9–10
Repairs, improper, 45, 51
Reverse bead angle, 65
Reversion, 15, 21
Right side vs. left side, determining, 149–150
Rim fitments, 64*n*
Rim flange, 171, 172, 173
   bending of, 83, 85
Rim pinch, 122, 123*f*
RMA, 46
Road hazards *See* Impacts
Road rash *See* Abrasions
Rough texture, 18, 21
Rubber
   characteristics of, 89–91
   thick/thin, 19–20
Rubber Manufacturers Association (RMA), 10, 107
Run-flat damage, 119–124
Run-flat sequence, 203–207
Rust, 43–45, 53–55, 71–73

SAE 1561/1633, 7
Safety belts, 157
Salt corrosion, 42, 45, 53–55
Scaling, 56
Serial side (SS), 161, 165–166
Shearography, 2, 128
Shell rating scale, 165*n*, 209–210

Sidewalls
   chemical damage to, 124–125
   cracking in, 90, 126
   creasing in, 91
   damaged, 82
   detached, 50–51, 83, 84*f*, 87
   exterior, contact with road surface, 59–60
   fractured, 76–80
   indentations in, 126
   oval area in, 83
   ozone deterioration in, 88*f*
   scuffing from curbing, 145–146, 147
   separation of, 114, 117–118
   serial side, 161, 165–166
   spare tire marks on, 126, 128
   spot ozone damage of, 101*f*
   white, 91, 92*f*, 118, 126
Siping, 32, 33
Snags, 12
   identification of, 102–104
Spare tire marks, 126, 128*f*
Speed, high, 7–8
   identification of, 108–109
Speed rating, 7
Splice locations, openings at, 132–133
Squirming, 30
Standing wave effect, 108
Stop/start marks, 20–21
Storage, 11–12, 54, 72–73, 91, 100
   improper, identification of, 106–107
Strain, 90
Sunlight, exposure to, 91, 106
Suspension, poor maintenance of, 11

Tactile inspection, 165–169
Tears, 12
   bead, 51, 93–98
   circumferential, 50*f*, 101
   directional, 104*f*
   identification of, 102–104
   short, 15, 16*f*, 18, 21, 47

Tears *(continued)*
  wheel flange, 98
Temperature
  ambient, 8
  glass transition, 153
  internal, 7
  pavement, 8
Tensile wire breaks, 27, 28*f*, 87, 108, 117*f*
Thick/thin rubber, 19–20
Tire and Rim Association (TRA), 6, 64*n*
Tire Industry Association (TIA), 107
Tires
  aging of, 9–10
  building process for, 154
  demounting from wheel, 169–170
  location on vehicle, 145–150
  maintenance of, 10–11
  matching wheel to, 174–175
  medium/heavy truck, 65*n*
Torque cracks, 123
Tourniquets, 157
Transition zone, 20
Tread
  attached to casing, 67–70
  and belt detachment, 15–23, 70–73
  chemical damage to, 124–125
  cracking of, 23, 25*f*
  creasing of, 66, 67*f*
  cutting and chipping of, 104–106
  damaged, 82
  depth of, 168
  fractured, 76–80
  inspection of, 60–61, 167–168
  irregular wear of, 108
  odd appearance of, 26
  separations of, 111–112

Valve
  defective, 122
  location of, 170
Vibration, 5–6
Visual inspection, 165–169

Water corrosion, 42, 43, 45, 53–55
Wear
  abnormal, 110
  accelerated 23, 24*f*, 29–33
  irregular, 31, 108
Weather checking *See* Ozone deterioration
Wheel weight clip marks, 140–142, 171, 174, 176
  vs. balance weight marks, 137
  depth of, 61–62
Wheels
  damaged, 122
  demounting tire from, 169–170
  matching to tire, 174–175
  flange damage, 95
  inspection of, 170–173
  taper angle of, 64–65
Whitewalls, 91, 92*f*, 118, 126

"X" pattern, 76, 78, 150
X-rays, 54–55, 73, 129

Under-inflation, 6
U.S. Department of Transportation
  (DOT), 2, 158

About the Author

Thomas Giapponi received his B.S.E. from Purdue University in 1976. He is a registered Professional Engineer (Connecticut) with more than 30 years of experience in the tire industry. Mr. Giapponi's professional career spans the Armstrong Rubber Company and Pirelli Tire North America in the capacities of Tire Design Engineer; Manager of Medium/Heavy Truck Tires; Manager of Light Truck and Passenger Car Tire Engineering; Director of Tire Engineering; and Director of Tire Testing. At Pirelli's tire manufacturing plant in Hanford, California, Mr. Giapponi directed Pirelli's North American research and development and quality as the Director of R&D, Technical Director, and the Director of Quality in the market, respectively.

In 2001, Mr. Giapponi started TRGtech Tire Consulting LLC (www.tireconsultant.com). The company performs tire forensic investigations and tire patent defense, in addition to tire and tire forensic analysis training and consulting.

Mr. Giapponi also is a member of SAE International and the Tire Society, and he is Past President of the Tire and Rim Association.